XVI Brazilian Colloquium on Orbital Dynamics 2012

Journal of Physics: Conference Series Volume 465

Serra Negra, Sao Paulo, Brazil
26-30 November 2012

ISBN: 978-1-62993-284-2
ISSN: 1742-6588

Printed from e-media with permission by:

Curran Associates, Inc.
57 Morehouse Lane
Red Hook, NY 12571

Some format issues inherent in the e-media version may also appear in this print version.

Printed by Curran Associates, Inc. (2013)

Institute of Physics
Dirac House, Temple Back
Bristol BS1 6BE UK

Phone: 44 1 17 929 7481
Fax: 44 1 17 920 0979

techtracking@iop.org

Additional copies of this publication are available from:

Curran Associates, Inc.
57 Morehouse Lane
Red Hook, NY 12571 USA
Phone: 845-758-0400
Fax: 845-758-2634
Email: curran@proceedings.com
Web: www.proceedings.com

XVI Brazilian Colloquium on Orbital Dynamics 2012

Journal of Physics: Conference Series Volume 465

Serra Negra, Sao Paulo, Brazil
26-30 November 2012

ISBN: 978-1-62993-284-2
ISSN: 1742-6588

IOP Institute of Physics

TABLE OF CONTENTS

PAPERS

Journal of Physics: Conference Series 465 (2013) 011001 doi:10.1088/1742-6596/465/1/011001

EDITORIAL

The XVI Brazilian Colloquium on Orbital Dynamics was held from 26–30 November 2012, at the Biazi Grand Hotel, Serra Negra, São Paulo, Brazil. The Brazilian Colloquia on Orbital Dynamics are scientific events that occur bi-annually and are designed to develop those areas of research in celestial mechanics, orbital dynamics, planetary science, fundamental astronomy, aerospace engineering, and nonlinear systems and chaos. The meeting has been held for 30 years and it brings together researchers, professors and students from South American and also from other continents.

Acknowledgements
National Council for Scientific and Technological Development – CNPq
Coordination for the Improvement of Higher Level – CAPES
São Paulo Research Foundation - FAPESP

Journal of Physics: Conference Series **465** (2013) 011002 doi:10.1088/1742-6596/465/1/011002

Peer review statement

All papers published in this volume of *Journal of Physics: Conference Series* have been peer reviewed through processes administered by the proceedings Editors. Reviews were conducted by expert referees to the professional and scientifi standards expected of a proceedings journal published by IOP Publishing.

Journal of Physics: Conference Series **465** (2013) 012001 doi:10.1088/1742-6596/465/1/012001

Powered Swing-By Maneuvers around the Moon

A F Silva[1], A F B A Prado[1] and O C Winter[2]

[1] Instituto Nacional de Pesquisas Espaciais, INPE, São José dos Campos, Brazil
[2] Grupo de Dinâmica Orbital & Planetologia, Univ. Estadual Paulista, UNESP, Guaratinguetá, Brazil

E-mail: aleferrazsilva@hotmail.com, prado@dem.inpe.br, ocwinter@gmail.com

Abstract. A Swing-By maneuver occurs when a satellite approaches a celestial body to gain or lose energy from its gravitational field. The present work studies Swing-By maneuvers that are combined with the use of an impulsive thrust in different directions during the passage of the spacecraft by the periapsis of its trajectory around the Moon. The main objective of this type of maneuver is the fuel economy for orbital transfers. From the results, it is visible that the best direction to apply the impulse is not the direction to the motion of the spacecraft, as might be expected. In fact, by using a different direction, it is possible to maximize the effects of the Swing-By by decreasing the periapsis distance and/or increasing the turning angle of the maneuver, that are the key parameters to specify the variation of energy due to the Swing-By. The changes in the periapsis distance and turning angle cause modifications in the geometry of the original Swing-By, generating a maneuver with new parameters. This new Swing-By compensates for the loss of energy transfer that results of applying the impulse in a in a non-tangential direction.

1. Introduction

The maneuver that makes a spacecraft to gain or lose energy from the gravity field of a celestial body due to a close approach is called Swing-By. Several space missions used this technique and there are many papers in the literature studying this type of problem. An interesting one is written by Longuski and Strange [1], which developed a graphical method to study Swing-By trajectories with the goal of reducing the mass when starting the mission and/or the flight time. The literature also shows applications of successive Swing-Bys, like in Dunham and Davis [2], that studied multiple swing-bys around the Moon. Prado and Broucke [3] simulated and classified Swing-By maneuvers using the planet Jupiter as the body for the close approach and calculated the effects of these Swing-Bys. The goal was to check which ones of those orbits have a passage near the Earth. It was shown which one of those trajectories have a potential use for missions involving departures and/or arrivals in the planet Earth. Prado [4] studied the Swing-By trajectories under the model given by the elliptic restricted three-body problem and classified them in four groups: elliptic direct, elliptic retrograde, hyperbolic direct and hyperbolic retrograde. Prado [4] also showed the effects of the eccentricity of the primaries and of the position of the secondary body in the results. Several families of orbits have been found. The results obtained can be used to find optimal parameters that solve several types of problems. McConaghy et al [5] combines Swing-Bys maneuvers with low thrust, in order to obtain optimal trajectories. Gomes and Prado [6] extended this type of study to cover the situation where the Swing-

Journal of Physics: Conference Series **465** (2013) 012001 doi:10.1088/1742-6596/465/1/012001

By is made by a cloud of particles, instead of a single one, that can happen when an explosion of a spacecraft occurs. Other examples of missions can be seen in references [7] and [8].

This present research studies the Swing-By maneuver combined with the application of an impulse, which is applied exactly when the spacecraft passes by the periapsis of its orbit around the Moon. This impulse can be applied in different directions. These conditions were studied in the Earth-Moon-Spacecraft system. This work is a generalization of the paper by Prado [9], where similar studies were made. The idea is to find maneuvers that can be later used by the ASTER mission [10], which is a mission planned to visit a triple Near-Earth Asteroid (NEA) called 2001 SN263.

There are three parameters that describe the Swing-By maneuver: V_∞, the magnitude of the approach velocity of the spacecraft with respect to M_2, that is the Moon; $\mathbf{r_p}$, the periapsis distance for the Swing-By, which is the vector connecting the Moon to the spacecraft; and ψ, that is the approach angle, the angle between the vector $\mathbf{r_p}$ and the line M_1-M_2 (Earth-Moon). The vectors are represented in bold.

For the case of the powered Swing-By, that is a Swing-By maneuver combined with the application of an impulse, there are two additional parameters to define the impulse: δV, the magnitude of the impulse, given in km/s and α, the angle between the velocity vector of the satellite in the first orbit and the impulse applied. This angle α defines the direction of the impulse applied.

2. Dynamical System

In this work we used the model given by the restricted three-body problem. Numerical integrations were performed to obtain the results. M_1, the Earth, is the primary body, with the largest mass, M_2, the Moon, is the secondary body, orbiting M_1, and M_3 is the particle, with negligible mass, that orbits the Earth and makes a Swing-By with the Moon. The canonical system of units is used, adopting the unit of distance as being the distance between the Earth and the Moon, the unit of mass is defined as the mass of the Earth added to the mass of the Moon and the unit of time is chosen such that the period of the motion of the primaries is 2π. In this way, the equations of motion in the rotational system (in this system M_1 and M_2 are in fixed positions) are [11]:

$$\ddot{x} - 2\dot{y} = \frac{\partial \Omega}{\partial x} \tag{1}$$

$$\ddot{y} + 2\dot{x} = \frac{\partial \Omega}{\partial y} \tag{2}$$

$$\Omega = \frac{1}{2}(x^2 + y^2) + \frac{1-\mu}{r_1} + \frac{\mu}{r_2} \tag{3}$$

The potential Ω, that includes the gravitational and the centrifugal forces, depends of r_1 and r_2, which are the distances Earth-spacecraft and Moon-spacecraft, respectively, in canonical units. It is clear that, when r_1 and r_2 are small or near zero, the numerical integration has accuracy problems. To avoid this situation the method given by the Lemaître Regularization [12] is used. This method avoids singularities by using a substitution of variables.

We call θ the true anomaly of the point where the impulse is applied, and, in general, it is obtained through the scalar product of $\mathbf{r_p}$ and $\mathbf{r}$, where $\mathbf{r}$ is the position vector of this point with respect to M_2. In this case, the study was done with the impulse applied at the periapsis, so $\mathbf{r_p}$ and $\mathbf{r}$ are coincident ($\mathbf{r} = \mathbf{r_p}$), then $\theta = 0°$ in Figure 1, that shows the geometry of the problem.

The steps of the algorithm to solve this problem are: 1) From the point for the application of the impulse, which in this case is $\theta = 0$ °, integrates the orbit backwards in time [13] to get the information of the first orbit, before the close approach; 2) At the point P, shown in Figure 1, the impulse is applied, with magnitude δV and making a direction α with respect to the velocity of the spacecraft; 3) From the application of the impulse, the integration of the orbit is performed forward in time to get the information from the new orbit, after the passage; 4) To finalize, the calculation of the variation of energy is made, by making the difference between the energies of the orbits after and before the close approach.

It is important to emphasize that the impulse is applied inside of the Moons' sphere of influence [14] and that the direction of this impulse is defined by the angle α, with respect to the velocity of the spacecraft. Then, by varying the values of the angle α, it is possible to find the geometry that leads to the maximum variation of energy.

3. Results

For all cases studied in this paper we used, as initial condition, $r_p = 1.1$ radius of the Moon. The goal is to get the best direction to apply the impulse, in order to have the maximum variation of energy.

We studied the cases where $\theta = 0°$, α varying from -180° to 180°, in steps of 0.1. The values that were used for ψ, are $\psi = 90°$, $\psi = 180°$, $\psi = 225°$, $\psi = 270°$ and $\psi = 315°$.

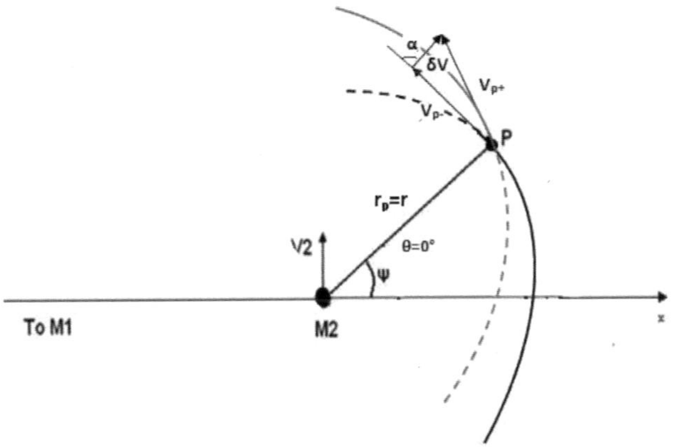

Figure 1 – Geometry of the Swing-By when applying an impulse at the periapsis of the orbit.

For the cases where $\psi = 90°$, the maximum variation of energy occurred when $\alpha = 0°$, for all values of the magnitude of the impulse. This fact happens because, when $\psi = 90°$, the maneuver has its maximum energy loss due to the Swing-By (see reference [9]). This conclusion comes from Equation (4) [9], that show the energy variation due to the Swing-By, where V_2 is the velocity of the Moon around the Earth and δ is half of the deflection angle.

$$\Delta E = -2V_2 V_\infty \sin(\delta)\sin(\psi) \tag{4}$$

In this situation it is necessary to avoid directions of impulses that can lower the periapsis distance, so increasing the effects of the Swing-By. It is necessary to take into account that, when applying the impulse in the vehicle, the magnitude of r_p may change. If α is negative, r_p tends to decrease, because the spacecraft is directed to the Moon. The point where the periapsis of the first orbit is located is also part of the second orbit, so it is impossible to reduce the value of r_p. Therefore, the best solution is to keep the original value of r_p, and the value of the angle α to reach this situation is zero. Figures 2 and 3 show the results for the case $\psi = 90°$. Those results are different from Prado [9], because this research considered the problem of the maximization of the magnitude of the variation of energy. In the preset work the magnitude is not taken and the variation of energy itself has to be maximized. So, in situations where the Swing-By decreases the energy of the spacecraft, the goal is to minimize this variation to reduce the loss. Those figures show the angle α, in degrees, in the horizontal axis and the variation of energy, in dimensionless units, in the vertical axis. It is visible the presence of a maximum variation for the energy.

Figures 4 and 5 show the results for the case $\psi = 270°$. In these cases, the maximum variation of energy occurred when α is negative. According to Prado [9], this fact happens because there is an increase in the deflection angle. Since the value of α is negative, the radial velocity gets a negative component from the impulse, and so the spacecraft is directed to the Moon. This makes the turning angle to increase and, consequently, the variation of energy is increased. Another reason to increase the performance of the Swing-By is that the periapsis distance is also decreased by the impulse with negative values of α, so a closer passage by the Moon is reached. There is a loss of energy due to the fact that the impulse is not applied in the direction of the motion of the spacecraft, but the net result of these gains and losses is positive, and then the gain due to the more efficient Swing-By compensates the loss of energy transfer that occurs due to the fact that the impulse vector is not aligned to the direction of the velocity vector of the spacecraft. This explanation covers all cases with $180° < \psi < 360°$. Table 1 shows more details, emphasizing the values of α that result in maximum variations of energy for the cases where $\psi = 225°$, $\psi = 270°$ and $\psi = 315°$. The symbol ΔE_{max} represents the maximum variation of energy and it is shown in the canonical system of units.

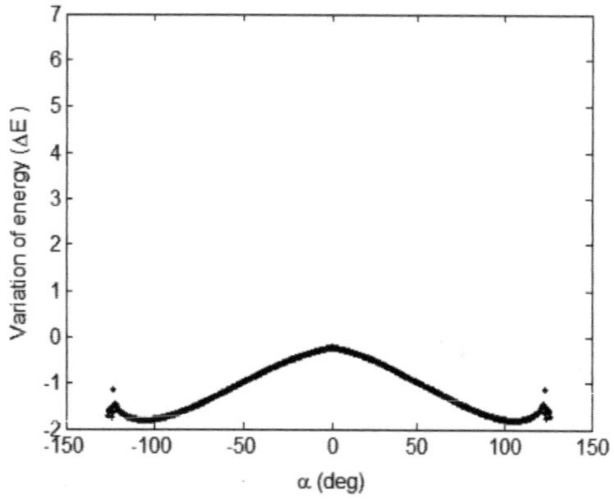

Figure 2 – Variation of energy (canonical system of units) for $\theta = 0°$, $\psi = 90°$ and $\delta V = 0.5$ km/s.

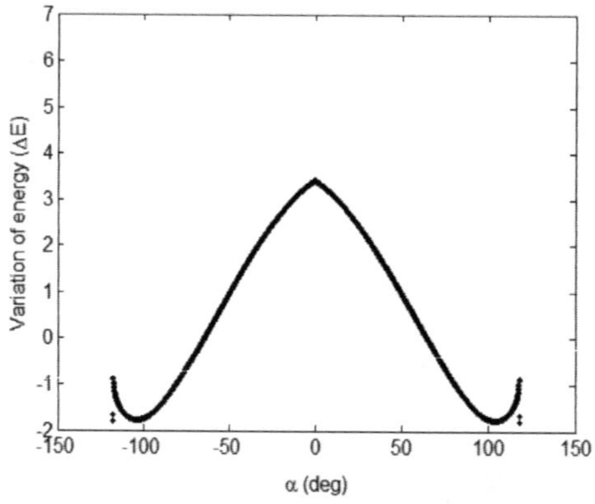

Figure 3 – Variation of energy (canonical system of units) for $\theta = 0°$, $\psi = 90°$ and $\delta V = 1.5$ km/s.

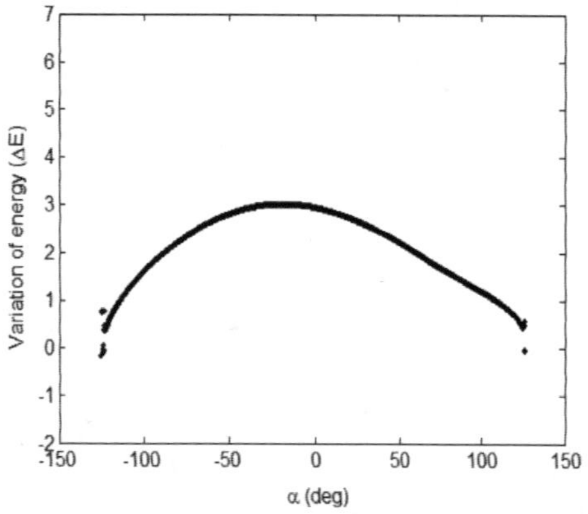

Figure 4 – Variation of energy (canonical system of units) for $\theta = 0°$, $\psi = 270°$ and $\delta V = 0.5$ km/s.

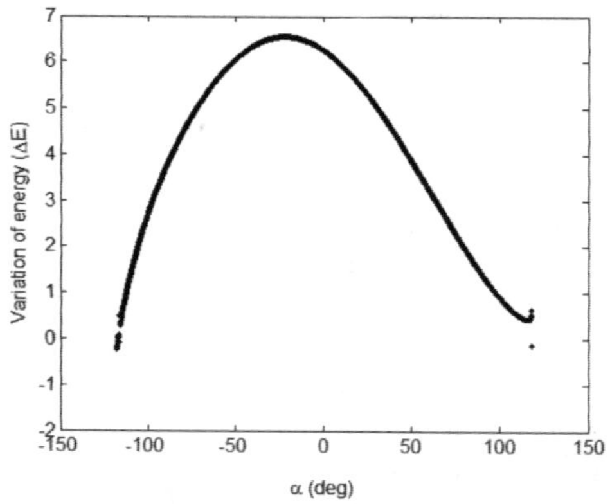

Figure 5 – Variation of energy (canonical system of units) for $\theta = 0°$, $\psi = 270°$ and $\delta V = 1.5$ km/s.

It is also visible from Table 1 that, when $\psi = 180°$, the maximum variation of energy occurs when α has a small negative value. This happens because when $\psi = 180°$, the effect of the Swing-By in the maneuver is null and, if the impulse has a negative component, the spacecraft is deviated to get closer to the Moon, so generating a trajectory that has an angle of approach greater than 180°, so increasing the energy of the spacecraft. At the same time the periapsis distance is reduced, what increases the gains due to the Swing-By. A larger deflection angle is also reached. Those facts added together generates an energy gain that would not exist with the application of the impulse at ψ = 180°.

Looking now at the results for ψ = 225°, ψ = 270° and ψ = 315°, it is visible that the value of α is always negative, for the reasons already explained. But the magnitude of the values are larger for ψ = 225°, because the orbits are still not located at the maximum value for the variation in energy, so the system has more flexibility to search for this optimal condition. After passing by the value ψ = 270°, the impulse is applied later in the passage by the periapsis, so there is less opportunity to influence the Swing-By. So the value of α is smaller, going close to zero, to take benefit of the larger transfer of energy due to the alignment of the impulse and the velocity of the spacecraft. The magnitude is smaller, but the value is still negative to cause the effects on the Swing-By already explained above.

Table 1. Maximum variation of energy for r_p = 1.1 Moon's radius.

θ	ψ	δV (km/s)	ΔE_{max}	α
0°	90°	2.0	5.6036	0°
0°	90°	2.5	8.0233	0°
0°	90°	3.0	10.6834	0°
0°	90°	3.5	13.5860	0°
0°	90°	4.0	16.7327	0°
0°	180°	2.0	3.8303	-6.5°
0°	180°	2.5	5.6045	-5.0°
0°	180°	3.0	7.6491	-4.0°
0°	180°	3.5	9.9574	-3.2°
0°	180°	4.0	12.5253	-2.7°
0°	225°	2.0	6.0703	-27.3°
0°	225°	2.5	8.0473	-26.2°
0°	225°	3.0	10.2934	-25.4°
0°	225°	3.5	12.8034	-24.8°
0°	225°	4.0	15.5739	-24.4°
0°	270°	2.0	8.6913	-22.0°
0°	270°	2.5	11.1007	-22.1°
0°	270°	3.0	13.7686	-22.1°
0°	270°	3.5	16.6932	-22.2°
0°	270°	4.0	19.8732	-22.2°
0°	315°	2.0	10.2319	-10.2°
0°	315°	2.5	13.0096	-10.7°
0°	315°	3.0	16.0312	-11.1°
0°	315°	3.5	19.2992	-11.5°
0°	315°	4.0	22.8151	-11.6°

4. Conclusions

With this study it was possible to analyze the importance of the direction of the impulse applied in a powered Swing-By maneuver, which is a maneuver that combines a close approach to a celestial body with the application of an impulsive thrust. The goal was to search for geometries that allow a maximum variation of the energy. The angle α helps to determine the curvature of the orbit of the spacecraft after the impulse, so reducing r_p and increasing the effects of the Swing-By. It was confirmed that the direction of the impulse is an important parameter in the powered Swing-By and it has strong influences in the behavior of the spacecraft. The results obtained also showed that, in most cases, captures occurred for values of α near the borders. That fact happened because when α is near -180° or 180°, there is a component of the impulse opposite to the motion of the spacecraft, that makes it to brake, thus reducing the size of the orbit and increasing the chance of capture.

The main conclusion is that the best way to apply the impulse in the orbit of the spacecraft is when it is applied out of the direction of the motion of the spacecraft, i.e., $\alpha \neq 0°$, usually making a negative angle and having a magnitude that depends on the angle of approach. The goal is to deviate the trajectory of the spacecraft to send it to pass closer to the Moon. This magnitude decreases with the angle of approach, been around 20 degrees for the situation where $\psi = 270°$. The reason for this decrease is that the spacecraft changes more its trajectory when it still did not pass by the geometry of maximum variation and so it tries to reach this situation. After passing by that, it changes its trajectory in smaller variation. When this situation happens, the maneuver gets less energy from the impulse, but it gains more energy from the Swing-By.

Acknowledgment
The authors wish to express their appreciation for the support provided by grants # 473387/2012-3 and 304700/2009-6, from the National Council for Scientific and Technological Development (CNPq); grants # 2011/09310-7 and 2011/08171-3, from São Paulo Research Foundation (FAPESP) and the financial support from the National Council for the Improvement of Higher Education (CAPES).

References

[1] Longuski J M and Strange N J 2002 Graphical method for gravity-assist trajectory Design *J. of Spac. and Rock* **39** 9-16
[2] Dunham D and Davis S 1985 Optimization of a multiple lunar-Swing-By trajectory sequence *J. of Astr. Sciences* **33** 275-288
[3] Prado A F B A and Broucke R A 1993 Jupiter Swing-By trajectories passing near the Earth *Adv. Astr. Sciences* **82** 1159-76
[4] Prado A F B A 1997 Close-approach trajectories in the elliptic restricted problem *J. of Guid. Control and Dyn* **20** 797-802
[5] McConaghy T T, Debban T J, Petropulos A E and Longuski J M 2003 Design and optimization of low-thrust gravity trajectories with gravity assist *J. of Spac. and Rock* **40** 380-387
[6] Gomes V M and Prado A F B A 2008 Swing-By maneuvers for a cloud of particles with planets of the Solar system *WSEAS Trans. on Appl. and Theor. Mech.* **3** 859-868
[7] Nasa LCROSS – Lunar crater observation and sensing satellite – LCROSS Overview. Page Editor: Robert Garner. 2010. Available in: <http://www.nasa.gov/mission_pages/LCROSS/overview/index.html> Ac.: 13 Mar. 2013.
[8] Nasa Messenger Trajectory Developed by JHU/APL 1999-2012. Available in: http://messenger.jhuapl.edu/the_mission. Accessed: 13 Mar. 2013.
[9] Prado A F B A Powered Swing-By 1996 *J. of Guid. Control and Dyn.* **19** 1142-47
[10] Sukhanov A, Velho H F C, Macau E and Winter O 2010 The Aster project: flight to a Near-Earth Asteroid *Cosm. Res.* **48** 443-450
[11] Murray C D and Dermott S F 1999 *Solar System Dynamics* (Cambridge: Cambridge Univ. Press)
[12] Lapa, L.C.; Regularização de Singularidades em Mecânica Celeste. 2008. f. 80. Dissertação (Mestrado em Matemática) - Universidade Federal de Pernambuco. Recife. 2008.
[13] Vieira Neto E and Winter O C 2001 Time analysis for temporary gravitational capture: satellites of uranus *Astron. J.* **122** 440
[14] Araujo R A N, Winter O C, Prado A F B A and Martins R V 2008 Sphere of influence and gravitational capture radius: a dynamical approach *Mont. Not. R. Astr. Soc.* **391** 675-684

Journal of Physics: Conference Series **465** (2013) 012002 doi:10.1088/1742-6596/465/1/012002

A Study of Swing-By Trajectories in the Galilean Satellites of Jupiter

A F B A Prado
Space Mechanics and Control Division, National Institute for Space Research -INPE, São José dos Campos-SP, BR.

E-mail: prado@dem.inpe.br

Abstract. The Swing-By maneuver is a technique that has been under use in space missions for many years now, with the goal of reducing the fuel consumption of a spacecraft travelling near a massive body and that requires a modification in its orbit. Several well-known missions used this approach, like Voyager, Galileo, Cassini, Ulysses, and many others. The present research has the goal of studying and classifying Swing-by maneuvers around the Galilean satellites of Jupiter. The approach to solve this problem is to simulate a large variety of initial conditions and then to perform a classification of the orbits according to the effects that the close approach made in the orbit of the spacecraft, in particular looking for captures and escapes that resulted from the close approach.

1. Introduction

The Swing-By is a very important tool used by several missions to save fuel consumption in orbital maneuvers. Several papers studied this problem, as can be seen in references [1] to [21]. In particular, examples of real missions that used this technique can be found in Minovich [6], that studied the trajectories for the Voyager mission, Farquhar and Dunham [4], that obtained trajectories to explore the Earth´s geomagnetic tail; Farquhar et. al. [5], that considered the orbital maneuvers for the ISEE-3/ICE comet mission; Dowling et. al. [3], that described several applications studied by JPL/NASA during the early years of 1962-1963; Weinstein [10], that studied a mission to Pluto and Swenson [9] that calculated trajectories to Neptune. Some other studies considered the Swing-by maneuvers under different aspects and models, like including an impulse during the passage [7]; considering elliptic orbits for the primaries [8]; comparing different dynamical models [12]; studying the close approach of a cloud of particles [13], [14], [15]; optimizing conditions for the maneuvers [16], [17]; obtaining orbits with successive passages [18], [19]; including the presence of an atmosphere in the planet [20] or finding trajectories to approach the Sun using the inner planets [21].

It is possible to use many different sets of initial conditions to identify one swing-by trajectory. In the present paper the variables used in Broucke [1] and Broucke and Prado ([2], [11]) are also used. They are: 1) J, the Jacobian constant of the spacecraft, that is an integral of the Restricted Three-Body Problem; 2) ψ, that is called "angle of approach" and is defined as the angle between the line of the periapsis of the orbit of the spacecraft around the Galilean satellite and the line connecting the primaries (Jupiter-Satellite) and; 3) R_p, the periapsis distance of the orbit of the spacecraft around the satellite.

So, by varying those variables, a numerical integration of the equations of motion is performed backward and forward in time, until the spacecraft is at a distance that can be considered far enough from the Galilean satellite, and its effects can be neglected. At those points, two-body celestial mechanics can be used to obtain the energy and the angular momentum before and after the Swing-By. Then, following Broucke [1] and Prado and Broucke [2], sixteen classes of orbits are defined according to the changes of energy and angular momentum made by the Swing-By and the orbits are mapped.

2. Definition of the Problem

The system is composed by three bodies, all of them assumed to be points of mass: Jupiter, the Galilean satellite and the spacecraft, assumed to have a negligible mass. Jupiter and the Galilean satellite are considered to be in circular orbits around their center of mass.

Then, the problem is to study the variation of energy and angular momentum of the spacecraft with respect to Jupiter due to the Swing-By. After obtaining the results it is possible to classify the orbits. For this classification, the orbits are divided into four categories: elliptic direct (positive angular momentum and negative energy), elliptic retrograde (negative angular momentum and energy), hyperbolic direct (positive angular momentum and energy) and hyperbolic retrograde (negative angular momentum and positive energy).

To have a better understanding of this maneuver, Fig. 1 explains the geometry of the Swing-By. The spacecraft comes from the point A, crosses the horizontal axis (Jupiter-Galilean Satellite line), arrives at the point P (periapsis of the orbit) and then goes up to the point B. Points A and B are chosen such that they are far enough from the Galilean Satellite and the system Jupiter-Spacecraft can be considered a two-body problem. Two of the initial conditions are marked in this figure: R_p (periapsis distance) and ψ (angle of approach). The Jacobian constant (J) of the spacecraft is the third initial condition.

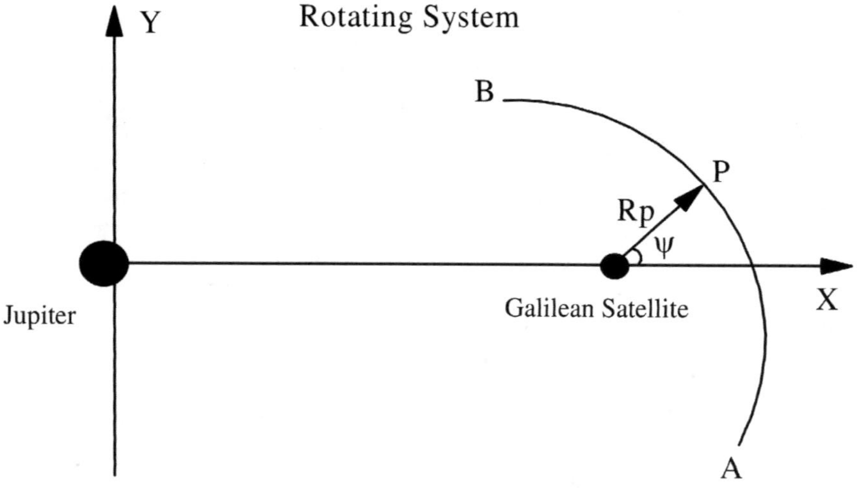

Fig. 1 - Geometry of the Swing-By.

3. Mathematical Model and Algorithm

The model used to describe the motion of the spacecraft is the planar restricted circular three-body problem. So, the equations of motion, in the usual canonical system of units for the spacecraft are:

$$\ddot{x}-2\dot{y}=x-\frac{\partial V}{\partial x}=\frac{\partial \Omega}{\partial x} \quad \ddot{y}+2\dot{x}=y-\frac{\partial V}{\partial y}=\frac{\partial \Omega}{\partial y} \tag{1-2}$$

$$\Omega=\frac{1}{2}\left(x^2+y^2\right)+\frac{(1-\mu)}{r_1}+\frac{\mu}{r_2} \tag{3}$$

To obtain the energy and the angular momentum of the spacecraft it is used [2]:

$$E=\frac{(x+\dot{y})^2+(\dot{x}-y)^2}{2}-\frac{1-\mu}{r_1}+\frac{\mu}{r_2}, \qquad C=x^2+y^2+x\dot{y}-y\dot{x} \tag{4-5}$$

To increase the accuracy in the numerical integrations, in particular when r_1 or r_2 are small, the Lemaitre regularization is used. It consists in a substitution of the physical variables (x, y, t) by another group of variables (ω_1, ω_2, τ), in such way that the singularities are eliminated. To perform this transformation, it is first defined a new complex variable q (q_1+i*q_2), with q_1 and q_2 given by:

$$q_1 = x + 1/2 - \mu \tag{6}$$
$$q_2 = y \tag{7}$$

After that, the transformation made is:

$$q = f(\omega) = \frac{1}{4}\left(\omega^2+\frac{1}{\omega^2}\right) \tag{8}$$

for the position variables (x-y) and:

$$\frac{\partial t}{\partial \tau} = |f'(\omega)|^2 = \frac{|\omega^4-1|^2}{4|\omega|^6} \tag{9}$$

for the time, where f '(ω) denotes $\frac{\partial f}{\partial \omega}$,

The equations of motion then become:

$$\omega''+2i|f'(\omega)|^2\omega' = \mathrm{grad}_\omega \Omega^* \tag{10}$$

with $\omega = \omega_1 + i*\omega_2$ is the new complex variable for position, ω' and ω" are the first and second derivatives of ω with respect to the regularized time τ; $\mathrm{grad}_\omega\Omega^*$ represents $\frac{\partial \Omega^*}{\partial \omega_1}+i\frac{\partial \Omega^*}{\partial \omega_2}$ and Ω^* is the transformed pseudo-potential given by:

$$\Omega^* = \left(\Omega - \frac{C}{2}\right)|f'(\omega)|^2 \tag{11}$$

where C = μ(1-μ) - 2J.

Then, a numerical algorithm is build to solve the problem. It has the following steps:

i) Numerical values are attributed to the three variables R_p, J, ψ;

ii) The initial position is obtained from $X_i = R_p\, cos(\psi) + (1-\mu)$, $Y_i = R_p sin(\psi)$ and the initial velocity from $V_{Xi} = -V sin(\psi)$, $V_{Yi} = +V\, cos(\psi)$, with $V = \sqrt{\dot{x}^2 + \dot{y}^2}$;

iii) A numerical integration of the equations of motion is performed forward in time until the spacecraft is far from the Galilean satellite. Then the energy (E_+) and angular momentum (C_+) are calculated;

iv) The initial conditions go back to the point P and the equations of motion are integrated backward in time. Energy (E-) and angular momentum (C-) are then obtained;

v) The variation in energy (E+ - E-) and angular momentum (C+ - C-) are then calculated.

4. Results

The results are shown in letter-plots, with one letter describing the outcome of the close approach. The figures have the Jacobian constant in the vertical axis and ψ in the horizontal axis. A figure is made for a fixed value of the parameter R_p. The letters A to P are defined in Table 1 and they are used for the classification of the orbits. Letter Z represents an orbit where the spacecraft stayed around the Galilean satellite for a long time.

Table 1 - Definition of the orbits

After: / Before:	Direct Ellipse	Retrograde Ellipse	Direct Hyperbola	Retrograde Hyperbola
Direct Ellipse	A	E	I	M
Retrograde Ellipse	B	F	J	N
Direct Hyperbola	C	G	K	O
Retrog. Hyperbola	D	H	L	P

The interval used here for the variables is: 180° ≤ ψ ≤ 360° and -1.35 ≤ J ≤ 1.55. The interval 0° ≤ ψ ≤ 180° is not used, because the figure is symmetric, since an orbit with $\psi = \theta$ differs from an orbit with $\psi = \theta + 180°$ by a time reversal. So, there is a correspondence: I⇔C, J⇔G, L⇔O, B⇔E, N⇔H, M⇔D. Orbits A, F, K and P are not changed.

For the periapsis distance the value 1.1 radius of the Galilean satellite is used, since it gives strong effects. Fig. 2 shows the results.

The numerical values for the limits involved in this research are:

1. Distance from the Galilean Satellite to the points A and B: 0.5 canonical units;
2. Distance for the spacecraft to be considered too far from Jupiter: 2.0 canonical units;
3. Mass of Jupiter = 1.89813 × 10^{27} kilograms;
4. Distance from Jupiter to the Sun: 778,330,001 km.

Table 2 - Data for the Galilean satellites in canonical units

Satellite	Mass	Radius
Io	0.00004725	0.004318
Europa	0.00002539	0.002326
Ganimedes	0.00007797	0.002458
Calisto	0.00005690	0.001280

Journal of Physics: Conference Series **465** (2013) 012002 doi:10.1088/1742-6596/465/1/012002

```
    1.55 PPPPPPPPPPPLKKKKKKKKKKKKKKKKKKKK
    1.45 PPPPPPPPPPPKKKKKKKKKKKKKKKKKKKKK
    1.35 FFPPPPPPPPPKKKKKKKKKKKKKKKKKKKKK
    1.25 FFFPPPPPPPPKKKKKKKKKKKKKKKKKKKKK
    1.15 FFFFPPPPPPPKKKKKKKKKKKKKKKKKKKKK
    1.05 FFFFFPPPPPPKKKKKKKKKKKKKKKKKKKKK
    0.95 FFFFFPPPPPPKKKKKKKKKKKKKKKKKKKKK
    0.85 FFFFFFPPPPPKKKKKKKKKKKKKKKKKKKKK
    0.75 FFFFFFPPPPPKKKKKKKKKKKKKKKKKKKKK
    0.65 FFFFFFFPPPPKKKKKKKKKKKKKKKKKKKKK
    0.55 FFFFFFFPPPPKKKKKKKKKKKKKKKKKKKKK
    0.45 FFFFFFFFPPKKKKKKKKKKKKKKKKKKKKKK
    0.35 ZFFFFFFFPPKKKKKKKKKKKKKKKKKKKKKK
    0.25 ZFFFFFFFNPKKKKKKKKKKKKKKKKKKKKKK
    0.15 ZFFFFFFFFPKKKKKKKKKKKKKKKKKKKKKK
J   0.05 ZFFFFFFFFPKKKKKKKKKKKKKKKKKKKKKK
   -0.05 ZFFFFFFFFBKKKKKKKKKKKKKKKKKKKKKK
   -0.15 ZFFFFFFFFAKKKKKKKKKKKKKKKKKKKKKK
   -0.25 ZFFFFFFFFAAKKKKKKKKKKKKKKKKKKKKK
   -0.35 ZFFFFFFFFAAKKKKKKKKKKKKKKKKKKKKK
   -0.45 ZFFFFFFFAAAAKKKKKKKKKKKKKKKKKKKK
   -0.55 ZFFFFFFFAAAAKKKKKKKKKKKKKKKKKKKK
   -0.65 ZFFFFFFAAAAAAKKKKKKKKKKKKKKKKKKK
   -0.75 ZFFFFFAAAAAAAIKKKKKKKKKKKKKKKKKK
   -0.85 ZFFFFAAAAAAAAAKKKKKKKKKKKKKKKKKK
   -0.95 ZFFBAAAAAAAAAAAKKKKKKKKKKKKKKKKK
   -1.05 ZZZZAAAAAAAAAAAAKKKKKKKKKKKKKKKK
   -1.15 ZZZZZZZZZZZZZZAAAKKKKKKKZZZZZZZZ
   -1.25 ZZZZZZZZZZZZZZZZZZZZZZZZZZZZZZZZ
   -1.35 ZZZZZZZZZZZZZZZZZZZZZZZZZZZZZZZZ
```

Io: $180° \leq \psi \leq 360°$

```
    1.55 PPPPPPPPPPPLKKKKKKKKKKKKKKKKKKKK
    1.45 PPPPPPPPPPPKKKKKKKKKKKKKKKKKKKKK
    1.35 FFPPPPPPPPPKKKKKKKKKKKKKKKKKKKKK
    1.25 FFFPPPPPPPPKKKKKKKKKKKKKKKKKKKKK
    1.15 FFFFPPPPPPPKKKKKKKKKKKKKKKKKKKKK
    1.05 FFFFFPPPPPPKKKKKKKKKKKKKKKKKKKKK
    0.95 FFFFFPPPPPPKKKKKKKKKKKKKKKKKKKKK
    0.85 FFFFFFPPPPPKKKKKKKKKKKKKKKKKKKKK
    0.75 ZFFFFFPPPPPKKKKKKKKKKKKKKKKKKKKK
    0.65 ZFFFFFFPPPPKKKKKKKKKKKKKKKKKKKKK
    0.55 ZFFFFFFPPPPKKKKKKKKKKKKKKKKKKKKK
    0.45 ZFFFFFFFPPKKKKKKKKKKKKKKKKKKKKKK
    0.35 ZFFFFFFFPPKKKKKKKKKKKKKKKKKKKKKK
    0.25 ZFFFFFFFNPKKKKKKKKKKKKKKKKKKKKKK
    0.15 ZFFFFFFFFPKKKKKKKKKKKKKKKKKKKKKK
J   0.05 ZFFFFFFFFPKKKKKKKKKKKKKKKKKKKKKK
   -0.05 ZFFFFFFFFBKKKKKKKKKKKKKKKKKKKKKK
   -0.15 ZFFFFFFFFAKKKKKKKKKKKKKKKKKKKKKK
   -0.25 ZFFFFFFFFAAKKKKKKKKKKKKKKKKKKKKK
   -0.35 ZFFFFFFFFAAKKKKKKKKKKKKKKKKKKKKK
   -0.45 ZFFFFFFFAAAAKKKKKKKKKKKKKKKKKKKK
   -0.55 ZFFFFFFFAAAAKKKKKKKKKKKKKKKKKKKK
   -0.65 ZFFFFFFAAAAAAKKKKKKKKKKKKKKKKKKK
   -0.75 ZFFFFFAAAAAAAIKKKKKKKKKKKKKKKKKK
   -0.85 ZZFFFAAAAAAAAAKKKKKKKKKKKKKKKKKK
   -0.95 ZZZZZZZAAAAAAAAAKKKKKKKKKKKKKZZZ
   -1.05 ZZZZZZZZZZZZZZZZZZZZZZZZZZZZZZZZ
   -1.15 ZZZZZZZZZZZZZZZZZZZZZZZZZZZZZZZZ
   -1.25 ZZZZZZZZZZZZZZZZZZZZZZZZZZZZZZZZ
   -1.35 ZZZZZZZZZZZZZZZZZZZZZZZZZZZZZZZZ
```

Europa: $180° \leq \psi \leq 360°$

```
    1.55 PPPPPPPPPPPLKKKKKKKKKKKKKKKKKKKK
    1.45 PPPPPPPPPPPKKKKKKKKKKKKKKKKKKKKK
    1.35 FFPPPPPPPPPKKKKKKKKKKKKKKKKKKKKK
    1.25 FFFPPPPPPPPKKKKKKKKKKKKKKKKKKKKK
    1.15 FFFFPPPPPPPKKKKKKKKKKKKKKKKKKKKK
    1.05 FFFFFPPPPPPKKKKKKKKKKKKKKKKKKKKK
    0.95 FFFFFPPPPPPKKKKKKKKKKKKKKKKKKKKK
    0.85 FFFFFFPPPPPKKKKKKKKKKKKKKKKKKKKK
    0.75 FFFFFFPPPPPKKKKKKKKKKKKKKKKKKKKK
    0.65 ZFFFFFFPPPPKKKKKKKKKKKKKKKKKKKKK
    0.55 ZFFFFFFPPPLKKKKKKKKKKKKKKKKKKKKK
    0.45 ZFFFFFFFPPKKKKKKKKKKKKKKKKKKKKKK
    0.35 ZFFFFFFFPPKKKKKKKKKKKKKKKKKKKKKK
    0.25 ZFFFFFFFNPKKKKKKKKKKKKKKKKKKKKKK
    0.15 ZFFFFFFFFPKKKKKKKKKKKKKKKKKKKKKK
J   0.05 ZFFFFFFFFPKKKKKKKKKKKKKKKKKKKKKK
   -0.05 ZFFFFFFFFBKKKKKKKKKKKKKKKKKKKKKK
   -0.15 ZFFFFFFFFAKKKKKKKKKKKKKKKKKKKKKK
   -0.25 ZFFFFFFFFAAKKKKKKKKKKKKKKKKKKKKK
   -0.35 ZFFFFFFFFAAKKKKKKKKKKKKKKKKKKKKK
   -0.45 ZFFFFFFFAAAAKKKKKKKKKKKKKKKKKKKK
   -0.55 ZFFFFFFFAAAAKKKKKKKKKKKKKKKKKKKK
   -0.65 ZFFFFFFAAAAAAKKKKKKKKKKKKKKKKKKK
   -0.75 ZFFFFFBAAAAAAIKKKKKKKKKKKKKKKKKK
   -0.85 ZZFFFBAAAAAAAAKKKKKKKKKKKKKKKKKK
   -0.95 ZZZZZZAAAAAAAAAKKKKKKKKKKKKKKKZZ
   -1.05 ZZZZZZZZZZZZZZZZZZZZZZZZZZZZZZZZ
   -1.15 ZZZZZZZZZZZZZZZZZZZZZZZZZZZZZZZZ
   -1.25 ZZZZZZZZZZZZZZZZZZZZZZZZZZZZZZZZ
   -1.35 ZZZZZZZZZZZZZZZZZZZZZZZZZZZZZZZZ
```

Ganimedes: $180° \leq \psi \leq 360°$

```
    1.55 PPPPPPPPPPPLKKKKKKKKKKKKKKKKKKKK
    1.45 PPPPPPPPPPPLKKKKKKKKKKKKKKKKKKKK
    1.35 FFPPPPPPPPPKKKKKKKKKKKKKKKKKKKKK
    1.25 FFFPPPPPPPPKKKKKKKKKKKKKKKKKKKKK
    1.15 FFFFPPPPPPPKKKKKKKKKKKKKKKKKKKKK
    1.05 ZFFFFPPPPPPKKKKKKKKKKKKKKKKKKKKK
    0.95 ZFFFFPPPPPPKKKKKKKKKKKKKKKKKKKKK
    0.85 ZFFFFFPPPPPKKKKKKKKKKKKKKKKKKKKK
    0.75 ZFFFFFPPPPPKKKKKKKKKKKKKKKKKKKKK
    0.65 ZFFFFFFPPPPKKKKKKKKKKKKKKKKKKKKK
    0.55 ZFFFFFFPPPLKKKKKKKKKKKKKKKKKKKKK
    0.45 ZFFFFFFFPPLKKKKKKKKKKKKKKKKKKKKK
    0.35 ZFFFFFFFPPKKKKKKKKKKKKKKKKKKKKKK
    0.25 ZFFFFFFFNPKKKKKKKKKKKKKKKKKKKKKK
    0.15 ZFFFFFFFFPKKKKKKKKKKKKKKKKKKKKKK
J   0.05 ZFFFFFFFFNKKKKKKKKKKKKKKKKKKKKKK
   -0.05 ZFFFFFFFFBKKKKKKKKKKKKKKKKKKKKKK
   -0.15 ZFFFFFFFFAKKKKKKKKKKKKKKKKKKKKKK
   -0.25 ZFFFFFFFFAAKKKKKKKKKKKKKKKKKKKKK
   -0.35 ZFFFFFFFBAAKKKKKKKKKKKKKKKKKKKKK
   -0.45 ZFFFFFFFAAAAKKKKKKKKKKKKKKKKKKKK
   -0.55 ZFFFFFFFAAAAIKKKKKKKKKKKKKKKKKKK
   -0.65 ZFFFFFFAAAAAAKKKKKKKKKKKKKKKKKKK
   -0.75 ZZZZZFBAAAAAAIKKKKKKKKKKKKKKKKKZ
   -0.85 ZZZZZZZZZZZAAAIKKKKKKKKKZZZZZZZZ
   -0.95 ZZZZZZZZZZZZZZZZZZZZZZZZZZZZZZZZ
   -1.05 ZZZZZZZZZZZZZZZZZZZZZZZZZZZZZZZZ
   -1.15 ZZZZZZZZZZZZZZZZZZZZZZZZZZZZZZZZ
   -1.25 ZZZZZZZZZZZZZZZZZZZZZZZZZZZZZZZZ
   -1.35 ZZZZZZZZZZZZZZZZZZZZZZZZZZZZZZZZ
```

Calisto: $180° \leq \psi \leq 360°$

Fig. 2 - Results for $R_p = 1.1$.

5. Conclusions

A numerical algorithm to calculate the effects of a close approach with the Galilean satellites of Jupiter in the trajectory of a spacecraft is developed. By varying the initial conditions of the trajectories, the orbits can be classified according to the changes in energy and angular momentum, to identify captures, escapes and situations where the sense of the orbit is reversed. It is shown that, although the Galilean Satellites have masses much smaller when compared to the giant planets, all of them can provide small regions of escape and capture trajectories from the Jupiter system, as well as some orbits that reverse the sense of the motion. A closer look at the plots made here reveals that all of them have similar characteristics. Those trajectories can be used to insert a spacecraft coming from the Earth in orbit around Jupiter, or to help a spacecraft to leave the Jupiter system.

Acknowledgment

The author wish to express their appreciation for the support provided by grants # 473387/2012-3 and 304700/2009-6, from the National Council for Scientific and Technological Development (CNPq); grants # 2011/09310-7 and 2011/08171-3, from São Paulo Research Foundation (FAPESP) and the financial support from the National Council for the Improvement of Higher Education (CAPES).

References

[1] Broucke R A 1988 The Celestial mechanics of gravity assist, *In:AIAA/AAS Astrodynamics Conf., Minneapolis*, 88
[2] Broucke R A and Prado A F B A 1993 *Adv. Astr. Sci.* **82**(II) 1159-1176
[3] Dowling R L, Kosmann W J, Minovitch M A. and Ridenoure R W 1991 Gravity Propulsion Research at UCLA and JPL *in: 41st Con. of the IAF* (Dresden, GDR, Oct. 1991) pp 6-12
[4] Farquhar R W and Dunham D W 1981 *Jour. Guid. Cont. and Dyn.* **4**(2) 192-196
[5] Farquhar R; Muhonen D; Church L C 1985 *Jour. Astr. Sci.* **33**(3) 235-254
[6] Minovich M A 1961 A Method for Determining Interplanetary Free-Fall Reconnaissance Trajectories *JPL Tec. Memo* 312-130 47 pp
[7] Prado A F B A 1996 *Jour. Guid. Cont. and Dyn.* **19**(5) 1142-47
[8] Prado A F B A 1997 *Jour. Guid. Cont. and Dyn.* **20**(4) 797-802
[9] Swenson B L 1992 Neptune Atmospheric Probe Mission *AIAA in: AIAA7AAS Astr. Conf.*
[10] Weinstein S S 1992 Pluto Flyby Mission Design Concepts for Very Small and Moderate Spacecraft *in: AIAA/AAS Astr. Conf.*
[11] Prado A F B A and Broucke R 2005 *Adv. Space Res.* **36**(3) 561-568
[12] Prado A F B A 2007 *Adv.Spac. Res.* **40** 113-117
[13] Formiga, J K S; Prado, A. F. B. A. 2011 Orbital characteristics of a cloud of particles performing a three dimensional swing-by. *Anais do DINCON, 2011*
[14] Gomes V M and Prado A F B A 2010 *WSEAS Trans. on Math.* **9** 811-820.
[15] Gomes V M and Prado A F B A 2008 *Trans. App. Theor. Mec.* **3** 11
[16] Strange N J and Longuski J M 2002 *Jour. Spac. and Roc.* **39**(1) 9-16
[17] McConaghy T T; Debban T J; Petropulos A E; Longunski J M 2003 *Jour. Spac. Roc.* **40**(3) 380-387
[18] Prado A F B A and Broucke R A 1996 *Jour. Guid. Cont. and Dyn.* **19**(4) 929-933
[19] Prado A F B A and Broucke R A 1995 *Jour. Guid. Cont. and Dyn.* **18**(3) 593-598
[20] Prado A F B A 1996 *Acta Astron.* **36**(6) 483
[21] Sukhanov A 1999 *Acta Astron.* **45** 177-185

Journal of Physics: Conference Series **465** (2013) 012003 doi:10.1088/1742-6596/465/1/012003

Topographic profile of a target with use of laser pulses. A survey directed to the Brazilian deep space mission ASTER.

A G V de Brum
Universidade Federal do ABC – UFABC, SP, Brazil
antonio.brum@ufabc.edu.br

A P Rodrigues
Universidade Federal do ABC – UFABC, SP, Brazil
ana.paula@aluno.ufabc.edu.br

Abstract: This work is directly related to the development of the laser altimeter for the ASTER mission, named ALR. The Brazilian deep space mission ASTER plans to send a small spacecraft to encounter and investigate the triple asteroid 2001-SN263. The launch is scheduled to occur in 2017 and the ALR is now under development in partnership with UNICAMP, UFABC and aerospace companies. In this work, the environment and the operation of the instrument were modeled and simulations were carried out in order to better understand and define the instrument parameters. The creation of the simulation software to control the operation of the instrument was the main purpose of this work, and the software so far created is the main result of it. The software was successfully tested with respect to some common expected situations.

Keywords: Laser altimeter, ALR, laser rangefinder, aerospace technology, Brazilian deep space mission, ASTER, asteroid 2001SN263.

1 – Introduction

Asteroids are primordial celestial bodies that can help scientists to understand the process of formation of the solar system. NEAs (Near Earth Asteroids) are especially interesting objects to investigate in space missions with robotic probes, since they periodically approach the Earth's orbit and eventually may threaten life on the planet.

In the first decade of the 21th century, a new era in space exploration began with the advent of spacecraft observations of asteroids. In 2000, the NEAR spacecraft went into orbit around the asteroid 433 Eros for a year of scientific investigation of its geology, mineralogy and surface chemistry [1, 2]. In 2005, the Hayabusa spacecraft arrived in its encounter with asteroid Itokawa for scientific observation of the asteroid shape, spin, topography, composition and density, and subsequent touch-down, sample collection and return to Earth [3, 4]. Both spacecrafts were equipped with a laser altimeter for navigation and topographical purposes [5]. Basically, such an onboard instrument is used to determine the elevation from the spacecraft to a given location on the surface of the asteroid by simply measuring the time delay between the transmission of a laser pulse and its return.

Space missions designed to investigate asteroids have become more important, mainly after the increasing interest in the mining of such bodies, where various minerals and water could be extracted. An example of a future mission to asteroids is the mission Hayabusa 2, whose possible launch window is in 2014 or 2015, with arrival in the asteroid named 1999JU3 in 2018, and return to Earth in 2020 [6]. In Brazil, laser altimetry for space missions is a new branch of research and development for engineers and scientists whose interest in this sector has grown as the country plans its first ever deep-space mission [7, 8].

Based on science and technical advisory reports conducted recently by Brazilian experts at universities, research centers and private companies, Brazil government has now recognized the importance of an ambitious deep-space mission in order to test key national technologies in space, to burst the national aerospace industry and also to stimulate the new generation of space scientists and engineers of the country.

2 – ASTER, A Brazilian Deep Space Mission

The ASTER mission is the first time Brazil is involved in a deep space mission. To explore the unknown triple near-Earth asteroid 2001-SN263, the mission will be undertaken in cooperation with Russia.

The ASTER mission spacecraft is based on a Russian MetNet platform. Its launch is scheduled to occur in 2017. The scientific instrumentation to fly will collect asteroid data for approximately 1 year and the main scientific objectives include determining the size, mass, volume, gravity field and rotation of the triple asteroid members, together with the identification of the composition, morphology and topography of the surface of each body. An investigation of the system dynamics will be conducted to obtain evidences of the system formation.

UFABC participates in this mission with the development of two instruments that will be used in the investigation of the asteroids, a laser altimeter, named ALR, and an infrared spectrometer.

The ALR (Aster Laser Rangefinder) will be useful in the navigation of the probe, in the phase of the mission when the spacecraft shall be maneuvered to reach a closer approach to the asteroid, as well as in the investigation and mapping of the asteroid features to be conducted in the proximity phase [9].

A laser rangefinder is a tool that works from the emission of laser pulses toward a specific target and the reception of the return signal reflected by the target surface. The measurement of the pulse time of flight allows calculating accurately the distance between emitter and target. Additionally, when the expanded pulse illuminates a small area of the target surface, the return pulse brings the information about the topographical features of the surface illuminated by the pulse. The continuous emission of pulses, combined with the probe-target movement, enables the evaluation of the whole surface, and by interpolation, the characteristic topography of the investigated body can be determined. The investigation of the target is conducted track by track. Reviews of data on tracks result in a topographic model of the target revealing its surface characteristics. Interpolations are used to estimate the areas not covered by the pulses between tracks.

3 – Methodology: Principles and Operation Parameters of the ALR

The topographic details of the area illuminated by a laser pulse, in this case the asteroid 2001-SN263, are obtained through the analysis of the received waveform. The asteroid surface is normally irregular and there are some common topographic features expected to be detected. Among them, the four simplest ones are: cliff, flat surface, inclined surface and depression, which have been analyzed in the literature [10]. Although other types of surface terrains were also subject of modeling and simulation, in this work, only the cliff type is described in detail and has its results presented, because it has two different possible well known results that can be used to verify the modeling and simulation process.

It is important to note that the waveform of the pulse received by the detector depends on the waveform of the emitted pulse, the topographical features of the surface and the way this surface reflects the emitted light (a diffuse reflection model was used). On the other hand the intensity of the received pulse depends on the intensity of the emitted pulse, the beam divergence angle, the distance between the instrument and the asteroid, the absorption of the environment where the light propagates (called the environment transmissivity), the aperture of the detector telescope, and on the specific reflectance of the surface with respect to the incident laser frequency. The results reported here were obtained through modeling software, considering all these intervening variables.

When the laser pulse is emitted and the footprint covers the entire length of a cliff, there are two expected waveforms for the return pulse, which can be seen in Figure 1.

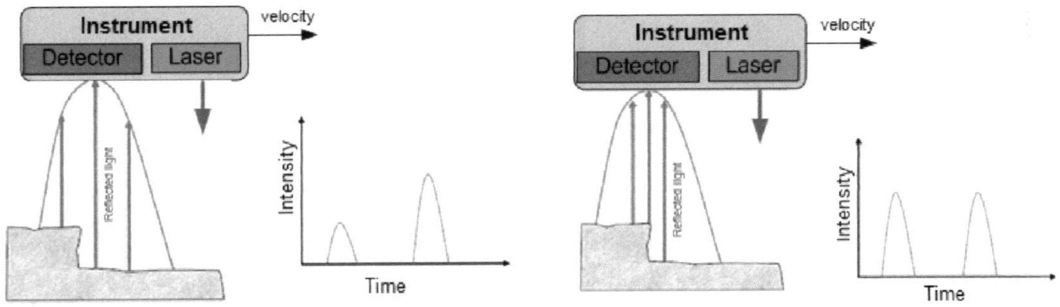

Figure 1: Types of cliff and the expected return pulses.

Source: Marti, Kilian. The BepiColombo Laser Altimeter - Detector Characterization and Implications for Mercury Science. Bern University. 2005.

If the pulse center is shifted toward a plain, the reflected pulse is split into two components with different intensities, because most of the intensity is emitted at the lower topography and take longer to be reflected.

If the center of the laser beam focus exactly at the point of variation in altitude between the surfaces of the cliff, the two components of the reflected pulse will have identical intensities because the intensities will also be distributed on the cliff, differing only about in the time of reflection.

4 – Results: Simulations and the ALR Control Software

The simulation of the return signal was performed with use of the MATLAB® environment [11]. The various parameters and modeling details will be the subject of another paper. It was expected to obtain the return waveforms found in the literature [10].

The use of the simulation software made it possible to simulate the emission of a laser pulse, the consequent target surface illumination (initially using the cylindrical "flat hat" spatial energy distribution model), the pulse reflection on the target surface (cliff), its return (after reflection) and detection with the registration of the return waveform. The simulation software was implemented considering the surface illuminated by the laser beam as a diffuse light-emitting surface, divided into four quadrants of multiple tiny light sources (Figure 2).

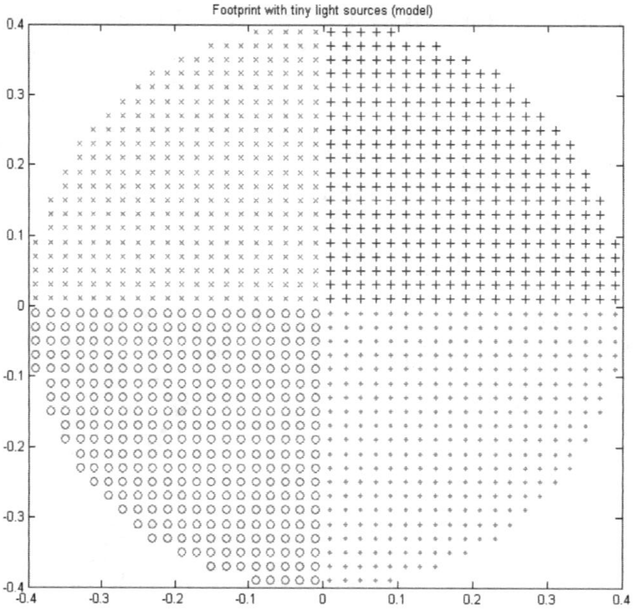

Figure 2: Surface of the target illuminated by the laser beam footprint, divided into four quadrants of multiple tiny light sources whose centers are shown. The axes in meters. In this simulation, each source has 2X2 cm^2 area, the divergence angle is 800 μrad and the distance target-instrument is 1,000 m.

In the implemented algorithm, the width of the laser pulse was standardized in 10 ns, a common pulse width found in the literature [5]. In general, the pulse width is identified at the half power of the emitted pulse (FWHM - *Full Wave Half Maximum*).

The specifications for the laser beam used in the simulation were taken from the literature [12], including those related to the ALR under development [9].

Initially, simulations were performed using a square wave as emitted pulse. Subsequently, the trapezoidal shape of the emitted pulse was used. This choice was taken because this second option better describes the actual case. Other shapes for the emitted pulse are also available and will be discussed in a future work (exponential and Gaussian).

The simulations described here were performed on surfaces shaped so as to present an altitude difference between two regions characterizing a cliff. The two situations shown in Figure 1 are considered. The altitude differences among the four quadrants (Figure 2) are manipulated by the user, thus changing the expected shape of the return wave.

One important parameter to be considered is the sampling time, i.e., the time interval used to integrate all energy reaching the detector sensor. This time is chosen according to the instrument objectives and is directly related to the desired vertical resolution. The use of a detector capable of sampling the return wave every 1 ns, as done here, means that the register of a full return waveform is wanted and that the vertical resolution of 15 cm was required. To provide such a sampling time, an oscillator of 1 GHz has to be used. The simulator developed is being used as a helpful tool in the analysis and choice of the pair detector-oscillator to fly in the ASTER mission. This choice, of course, has to take into consideration the best cost-benefit relation.

As one of its outputs, the simulator offers a graph of the detected return pulse. The results of the first simulation are show in Figure 3. In this case, half of the pulse energy reaches the relief with highest altitude (distance target-surface 1,000 m), and the other half reaches the surface with lower altitude (distance target-surface 1,003 m).

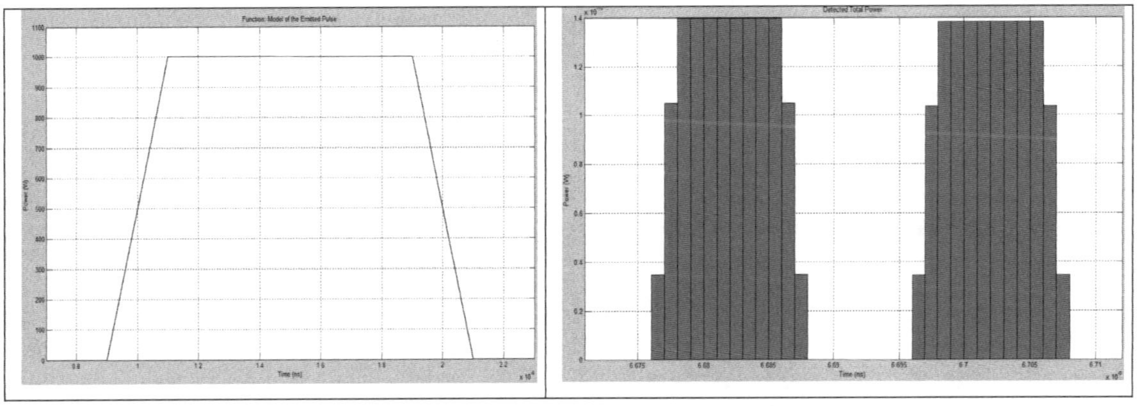

Figure 3: Emitted laser pulse function (trapezoidal; left), and its characteristic return wave (right).

In the first simulation, two waveforms are obtained in return with practically the same power, as shown in Figure 3 (right). The intensity of the return signal is very low, of the order of nW (10^{-9} Watts). This expected result is consistent with literature [12] and shows clearly the need of signal amplification. The two separate waves are due to the fact that the altitude difference between the two surfaces is bigger than half of the distance the light travels during the pulse emission time interval, for the parameters used. That is, when the photons reflected by the most distant surface return to the instrument detector, all photons reflected by the closest surface have already arrived and were detected nanoseconds before, causing the observed interval between pulses in the return waveform. An observation is made here about the slight difference in power intensity seen in Figure 3 (right) that is due mainly to the diffusion of the light coming from the reflection on the second surface (3 m more distant). The intensity of the light reaching the detector is inversely proportional to the square of the distance source-detector. Also, the environment absorption of the travelling light has been considered.

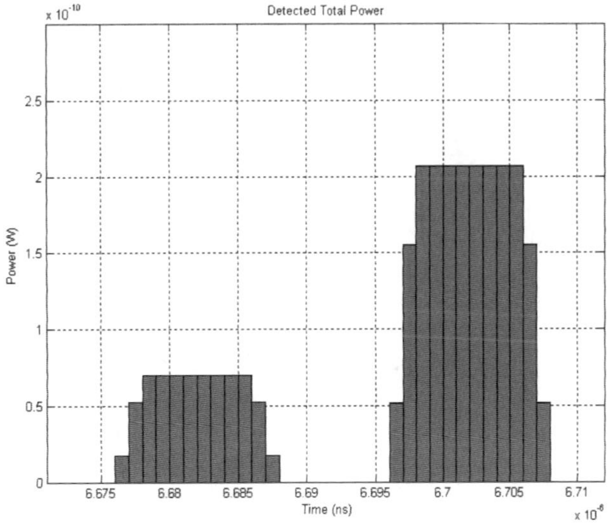

Figure 4: Return waveform of the second simulation. Obtained during the simulation on a surface Cliff with 30cm of difference between the surfaces. 1/4 of the pulse reaches the relief with highest altitude and 3/4 reaches the surface with lower altitude.

The results of the second simulation are shown in Figure 4. The emitted pulse is the same as in Figure 3. Here, 1/4 of the pulse reaches the relief with highest altitude and 3/4 reaches the surface with lower altitude. The graph obtained confirms the expected shapes, time intervals and power intensities, according to the cited literature, thus verifying the adequacy of the developed software.

5 – Conclusions

Several simulations were performed, initially starting with a square wave pulse and subsequently with use of a trapezoidal wave, which is closer to the real case and whose results were presented. In both cases, the simulations showed results very close to those expected in the literature, proving the functionality of the software.

The main motivation for the creation of the laser altimeter simulator, described partially here, is the development of software to control the operation of the ALR, the laser altimeter to fly in the Brazilian mission to deep space – ASTER.

The software simulator is still under development at UFABC. Many of its modules will be used in the control and processing unit that accompanies the apparatus as a whole. It is now being improved to contain more specific modules (different choices for the emitted pulse function, for instance), and to include the processing of the return wave pulse. Also it has been used as a tool to better define instrument parameters and parts.

6 – References

[1] Chapman, C. R., 1995, "Near Earth Asteroid Rendezvous: Eros as the key to the S-type conundrum", Proc. XXVI Lunar Planetary Science Conference, pp. 229-230.
[2] Cole, D. Timothy. NEAR Laser Rangefinder: A Tool for the Mapping and Topoligic Study of Asteroid 433 Eros. Johns Hopkings APL Technical Digest. v.19, n.2, 1998.
[3] Yano, H., et al., 2006, "Touchdown of the Hayabusa spacecraft at the Muses Sea on Itokawa", Science, Vol. 312, no. 5778, pp. 1350-1353
[4] Shinsuke Abe, et al. Mass and Local Topography Measurements of Itokawa by Hayabusa. Science. v.312, n.1344, June 2006. DOI: 10.1126/science.1126272
[5] Mukai, T., Abe, S., Barnouin-Jha, O., and Cheng, A., 2008, "Hayabusa LIDAR", NASA Planetary Data System, 03 Feb 2013 http://sbn.psi.edu/pds/resource/haylidar.html
[6] Yano, H., et al. Planetary Protection of Hayabusa-2 mission, a sample return from 1999 JU3, C-type NEO, May 1-2,2012. Presentation for the NASA Planetary Protection Subcommittee 2012. at: http://science.nasa.gov/media/medialibrary/2012/05/04/Hayabusa-2_--_Hajime_Yano.pdf
[7] Sukhanov, A. A., Velho, H. F. C., Macau, E. E. N. and Winter, O. C., 2010, "The Aster Project: Flight to a Near-Earth Asteroid", Cosmic Research, Vol. 48, pp. 433-450.
[8] Macau, Elbert E. N.; Winter, Othon C.; Velho, Haroldo F. Campos; Sukanov, Alexander A.; Brum, Antonio Gil V. de ; Ferreira, J. Leonardo; Hetem Jr., Annibal; Sandonato, G. M. ; Sfair, Rafael. The Aster Mission: Exploring for the First Time a Triple System Asteroid. In: Prodeedings of the 62nd International Astronautical Congress, 2011, v. 1. Cape Town, 2011.
[9] Brum, Antonio G. V. de ; Hetem Jr. , Annibal ; Rego, I. da Silveira ; Francisco, Cayo, P. F. ; Fenili, A. ; Madeira, Fernando ; Cruz, F. C. ; Assafin, Marcelo . Preliminary Development Plan of the ALR, the Laser Rangefinder for the ASTER Deep Space Mission to the 2001 SN263 Asteroid.. Journal of Aerospace Technology and Management, JATM, v. 3, p. 331-338, 2011. doi: 10.5028/jatm.2011.03033611.
[10] Marti, Kilian. The BepiColombo Laser Altimeter – Detector Characterization and Implications for Mercury Science. Bern University. 2005.
[11] http://www.mathworks.com/ (accessed on 09/03/2013).
[12] Shan, Jie e Toth, Charles K. Topographic Laser Ranging and Scanning: Principles and Processing. CRC Press: Boca Raton, 2009, 590 pp.

Preliminary Attitude Control Studies for the ASTER Mission

Bruno Victorino Sarli[1], André Luís da Silva[2], Pedro Paglione[3]

[1] Doctorate student, InstitutoTecnológico de Aeronáutica, ITA, São José dos Campos, Brazil
[2]Assistant professor, Universidade Federal do ABC, UFABC, Santo André, Brazil
[3]Professor, InstitutoTecnológico de Aeronáutica, ITA, São José dos Campos, Brazil

Email: bruno.ren@gmail.com, andreluis.silva@ufabc.edu.br, paglione@ita.br

Abstract. This work discusses an attitude control study for the ASTER mission, the first Brazilian mission to the deep space. The study is part of a larger scenario that is the development of optimal trajectories to navigate in the 2001 SN263 asteroid system, together with the generation of orbit and attitude controllers for autonomous operation. The spacecraft attitude is defined from the orientation of the body reference system to the Local Vertical Local Horizontal (LVLH) of a circular orbit around the Alpha asteroid. The rotational equations of motion involve the dynamic equations, where the three angular speeds are generated from a set of three reaction wheels and the gravitational torque. The rotational kinematics is represented in the Euler angles format. The controller is developed via the linear quadratic regulator approach with output feedback. It involves the generation of a stability augmentation (SAS) loop and a tracking outer loop, with a compensator of desired structure. It was chosen the feedback of the p, q and r angular speeds in the SAS, one for each reaction wheel. In the outer loop, it was chosen a proportional integral compensator. The parameters are tuned using a numerical minimization that represents a linear quadratic cost, with weightings in the tracking error and controls. Simulations are performed with the nonlinear model. For small angle manoeuvres, the linear results with reaction wheels or thrusters are reasonable, but, for larger manoeuvres, nonlinear control techniques shall be applied, for example, the sliding mode control.

1. Introduction

ASTER is the first Brazilian mission to the deep space. It will study the triple asteroid system 2001 SN263. The attitude control study presented here is part of a broader scenario of development of orbit and attitude controllers for autonomous operation. They involve the generation of optimal trajectories to navigate inside the asteroid system, which were already developed in [1]. The 2001 SN263 is composed of three bodies: Alpha (central), Gamma (inner orbit) and Beta (outer orbit) [2]. The objective of this paper is to develop a controller to stabilize and track attitude commands for a spacecraft in a circular orbit around Alpha, which suffers perturbations from Beta and Gamma. The ephemeris of Beta and Gamma around Alpha was previously determined using numerical integration, from physical information of the system given from former studies [3]. The rotational kinematics is represented in the Euler angles format. The controller is developed via the linear quadratic regulator approach with output feedback [4]; it involves the generation of a stability augmentation (SAS) loop, together with a tracker in an outer loop composed by a compensator of desired structure. The feedback of the angular speeds was chosen in the SAS, while a proportional integral compensator was set in the

Journal of Physics: Conference Series **465** (2013) 012004 doi:10.1088/1742-6596/465/1/012004

outer loop. The control parameters are tuned using a numerical minimization that represents a linear quadratic cost, with weighting in the tracking errors and controls [5]. The compensator design is performed with a linearized attitude model for the circular orbit around Alpha, but the simulations are performed with the nonlinear model. The possibilities of use reaction wheels and thrusters as actuators are compared.

2. 2001 SN263

The 2001 SN263 system was discovered by the Lincoln Near-Earth Asteroid Research (LINEAR) project in 2001. Further observations made on 2008 revealed that what was believed to be one single asteroid was in fact a system of asteroids. Observations made during 16 days on February 2008 by the radio astronomers of the Arecibo Observatory in Porto Rico, showed that the 2001 SN263 is in fact a triple system composed by a central body and two satellites (figure 1), Alpha, Beta and Gamma [2]. The components of the system have diameters of about 2.8 km (Alpha), 1.2 km (Beta) and 0.5 km (Gamma). The two smaller bodies orbit Alpha with a semi-major axis of about 17 km (period of 147hrs) and a semi-major axis of about 4 km (period of 46hrs) [3].

Figure 1. The asteroid triple system 2001 SN263. Sequence of images obtained on February 13, 14, 18, 21 and 23rd, 2008 [2].

The evolution of Beta and Gamma around Alpha were first studied at [3] and no official accurate model has been derived yet. In order to obtain the ephemeris for the period of the mission, the initial conditions extracted from [3] were propagated using an Adams integrator for 20 days from June 1st of 2019 until June 20th of the same year.

3. Equation of motion

Some of the spacecraft's characteristics relevant to this work are: an initial wet mass of 152 to 157 kg with 66 to 71 kg of propellant at the time of launch. In this sense, it is assumed that the spacecraft will reach the system with approximately 100 kg of total mass. Also, the consideration is made that the mass does not vary during the control scenario of this paper.

The model used for control is composed by attitude equations of motion of a rigid spacecraft (equation (1)) with three reaction wheels placed in the body's principal axis of inertia (equations (3), (4) and (5)), or with clusters of thrusters which can deliver a pure rotational acceleration at each of the principal axes (equations (7), (8) and (9)). Both models include perturbation originated from the gravity gradient torque of the three asteroids present in the system (equation (2)). The deduction of these equations is commonly found in the attitude dynamics literature.

The spacecraft angular velocity propagation is presented by equation (1); where, I represents the inertia tensor that, when referenced to the principal axes of inertia, results in a diagonal matrix I =

diag(I_1, I_2, I_3), ω is the angular velocity given by p, q and r in each axis, $\vec{H}$ is the angular momentum and $\vec{M}$, the external torque.

$$\frac{d\vec{H}}{dt} = I \cdot \dot{\vec{\omega}} + \vec{\omega} \times \vec{H} = \vec{M} \tag{1}$$

The system's gravity gradient torque is represented by equation (2); where, $c_{i,3}$ is the 3rd column of the transformation matrix from the inertial frame to the body-fixed frame, R_j is the distance between the spacecraft's centre of mass and the asteroid's centre of mass, j represents the number of the body: Alpha is 1, Beta 2 and Gamma 3.

$$\{g_c\}_\beta = \sum_{j=1}^{n} 3\left(\frac{GM_j}{R_j^{\,3}}\right) c_{i,3}{}^{\times} I c_{i,3} \rightarrow \begin{bmatrix} M_x \\ M_y \\ M_z \end{bmatrix} = \sum_{j=1}^{n} 3\frac{GM_j}{R_j^{\,3}} \begin{bmatrix} (I_3 - I_2)c_{23,j}c_{33,j} \\ (I_1 - I_3)c_{33,j}c_{13,j} \\ (I_2 - I_1)c_{13,j}c_{23,j} \end{bmatrix} \tag{2}$$

The equations (3) to (6) describe the spacecraft angular acceleration due to reaction wheels; where, h_i represents the momentum stored in each reaction wheel, J_a is the inertia of the reaction wheel (common to the three elements) and u_i is the torque of the wheel applied to the axis of the internal rotor, which is assumed to be equal to the temporal variation of the rotor's angular momentum.

$$\dot{p} = \frac{(-qr+M_x)(x-1)y+(h_3q-h_2r+u_1)(1+xy)}{-1+J_a+x+J_axy} \tag{3}$$

$$\dot{q} = \frac{(-pr+M_y)(x+y)+(-h_3p+h_1r+u_2)(1+xy)}{(J_a-1)(1+xy)} \tag{4}$$

$$\dot{r} = \frac{(pq-M_z)(1+y)x+(h_2p-h_1q+u_3)(1+xy)}{J_a(1+xy)-1-y} \tag{5}$$

$$x \triangleq \frac{I_2-I_1}{I_3};\ y \triangleq \frac{I_3-I_2}{I_1} \tag{6}$$

Finally, equations (7) to (9) present the spacecraft angular acceleration due to thrusters; where u_i is the external torque around applied in the axis i, they are the controlled total thrust force (two anti parallel vectors) that include the thrust force and distance from the spacecraft's centre of mass.

$$\dot{p} = y\left[-qr + M_x - \frac{u_1}{I_2}\right] \tag{7}$$

$$\dot{q} = \frac{x+y}{1+xy}\left[pr - M_y - \frac{u_2}{I_2}\right] \tag{8}$$

$$\dot{r} = x\left[-pq + M_z - \frac{u_3}{I_2}\right] \tag{9}$$

4. Control structure and design

The attitude control is based on two loops with distinct function, the first loop is responsible for the system stabilization based on the angular velocity (SAS) and the second loop is responsible for tracking the commands based on the attitude of the spacecraft in Euler angles. Figure 2 presents the diagram of the control structure. It depicts the two loops and their respective gains, as well as the attitude dynamics and the control input, u.

The linear quadratic regulator technique is used to compute the controllers' gains, as suggested in [4]. It involves the generation of a stability augmentation (SAS) loop with p, q and r feedback and a tracking outer loop, with a proportional integral (PI) compensator. The parameters are adjusted using a numerical minimization that represents a linear quadratic cost, with weighting matrixes in the tracking error and controls [4], [5].

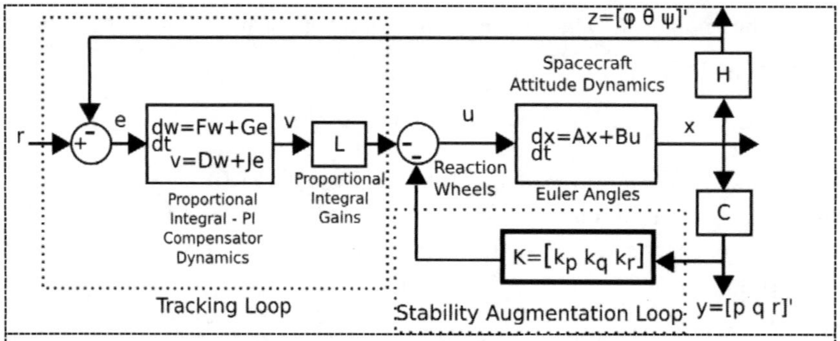

Figure 2. Control structure diagram, where the tracking loop is feed by the angular position in Euler angles and the stability augmentation loop feed by the angular velocity.

The dynamics for the plant and compensator are defined as [5]:

$$\dot{x}_a = A_a x_a + B_a u + G_a r, y_a = C_a x_a + F_a r, z = H_a x_a \quad (10)$$

$$x_a = \begin{bmatrix} x \\ w \end{bmatrix}, y_a = \begin{bmatrix} y \\ v \end{bmatrix} \quad (11)$$

$$A_a = \begin{bmatrix} A & 0 \\ -GH & F \end{bmatrix}, B_a = \begin{bmatrix} B \\ 0 \end{bmatrix}, G_a = \begin{bmatrix} 0 \\ G \end{bmatrix}, C_a = \begin{bmatrix} C & 0 \\ -JH & D \end{bmatrix}, F_a = \begin{bmatrix} 0 \\ J \end{bmatrix}, H_a = [H \quad 0] \quad (12)$$

The variables in the above equations are defined in figure (2). The control can be given by the augmented equation $u = -[K \quad L]\begin{bmatrix} y \\ v \end{bmatrix} = -K_a y_a$. The control gains are determined from the minimization of the quadratic index $J = \frac{1}{2}\int_0^\infty \tilde{e}^T\tilde{e} + \tilde{u}^T R \tilde{u}\, dt$; where, $\tilde{e} = e - \bar{e}, \tilde{u} = u - \bar{u}$ are the deviations of the tracking error and control input with respect to their steady state values $\bar{e}$ and $\bar{u}$. In reference [4], the problem of minimize the *functional* in the quadratic index J is converted to a problem of *function* minimization. In this sense, the performance index is given by $J = \frac{1}{2}tr(PX)$; where 'tr' is the trace operator and the matrices P and X are given by:

$$A_c^T P + P A_c + Q + C_a^T K_a^T R K_a C_a = 0 \quad (13)$$

$$A_c = A_a - B_a K_a C_a, B_c = G_a - B_a K_a F_a \quad (14)$$

In equation (14), A_c and B_c are the closed loop matrices. The equation (13) is of Lyapunov's type, it generates the symmetric positive definite matrix P. The existence and uniqueness of P requires that A_c is stable; a stable matrix in this context means that it is associated with a stable closed loop system, i.e. all their eigenvalues shall have negative real parts. The matrix Q is the weighting in the state, given from the output matrix H_a: $Q = H_a^T H_a$. The matrix X in the performance index is a quadratic function of the steady state $\bar{x}$: $X = \bar{x}\bar{x}^T = A_c^{-1} B_c r_0 r_0^T B_c^T A_c^{-T}$; where $A_c^{-T} = (A_c^T)^{-1}$ and r_0 is the amplitude of a step function applied as a reference command. Hence, the design is performed under the *hypothesis of step reference commands*.

In the design of the PI compensator, the following matrices where chosen:

$$F = \begin{bmatrix} 0 & 0 \\ 0 & 0 \end{bmatrix}, G = \begin{bmatrix} 1 & 0 & 0 \\ 0 & 1 & 0 \\ 0 & 0 & 1 \end{bmatrix}, D = \begin{bmatrix} 1 & 0 & 0 \\ 0 & 0 & 0 \\ 0 & 1 & 0 \\ 0 & 0 & 0 \\ 0 & 0 & 1 \\ 0 & 0 & 0 \end{bmatrix}, J = \begin{bmatrix} 0 & 0 & 0 \\ 1 & 0 & 0 \\ 0 & 0 & 0 \\ 0 & 1 & 0 \\ 0 & 0 & 0 \\ 0 & 0 & 1 \end{bmatrix} \quad (15)$$

$$K = \begin{bmatrix} k_p & 0 & 0 \\ 0 & k_q & 0 \\ 0 & 0 & k_r \end{bmatrix}, L = \begin{bmatrix} k_{\phi i} & k_{\phi p} & 0 & 0 & 0 & 0 \\ 0 & 0 & k_{\theta i} & k_{\theta p} & 0 & 0 \\ 0 & 0 & 0 & 0 & k_{\psi i} & k_{\psi p} \end{bmatrix} \quad (16)$$

The matrices in equation (15) define the *structure of the compensator,* and equation (16), in turn, determines a *predefined structure* for the control gain matrices. There, k_s is the gain of the angular speed $s = p, q, r$; k_{ai} and k_{ap} are the integral and proportional gains of the tracking error in the angle $a = \phi, \theta, \psi$.

During the design of the controllers for the spacecraft, the matrices of the linearized equations were obtained from the truncation of first order terms in the Taylor series expansions. In this process, the equations were also simplified by removing the terms associated with the gravity gradient torque of the satellites Beta and Gamma. In this way, the linear model treats these elements as *external perturbations*.

The design parameter is the control weighting matrix R. In the determination of the control laws for booth actuators, thrusters or reaction wheels, this matrix is $R = \frac{1}{100}\begin{bmatrix} 1 & 0 \\ 0 & 1 \end{bmatrix}$.

The numerical implementation is performed in MATLAB. The minimization of the cost $J = \frac{1}{2}tr(PX)$ is carried out by the function *fmincon*. The process that determines the control gains also requires an *initial stabilizing guess*. This is done by a previous minimization where the cost is the spectral radius of matrix A_c (the spectral radius of a matrix is the largest real part of their eingenvalues).

5. Results

This section present closed loop simulations performed with the nonlinear model also considering the gravity gradient torques of the Beta and Gamma. Additionally, the results are determined with perturbations represented by initial conditions in angular velocity, $[p_0, q_0, r_0] = [-2°/s, -3°/s, 5°/s]$, and angles, $[\phi_0, \theta_0, \psi_0] = [1°, -2°, 4°]$. Figure 3 shows the response to a tracking command given by a step function of amplitude $[\phi_r, \theta_r, \psi_r] = [2°, 1°, -10°]$.

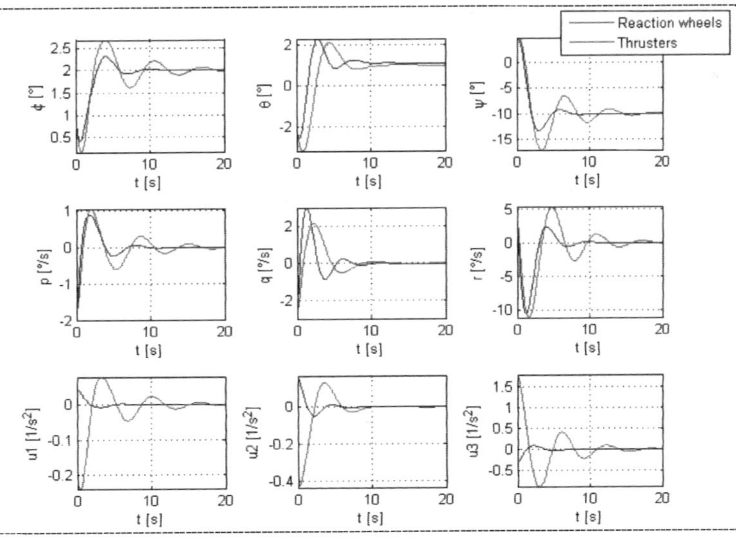

Figure 3. Closed loop response to step inputs on the Euler angles and perturbations in the angular velocity.

These results show that both actuators can cope with the disturbance; however, the torque generated by the thrusters has amplitudes with one order of magnitude smaller that the amplitudes for the reaction wheels torque; which is understandable since the acceleration provided by the thrusters can be much higher than the one provided by the reaction wheels. A better settling time can be noticed with the use of thrusters. They also generate responses with smaller overshoots. Nevertheless, the result given by using thrusters can also be reached with reaction wheels by properly adjusting the control weighting matrix. Moreover, the extra time gained with a faster convergence is negligible with respect to the orbital period. So, booth actuators are a reasonable choice for this task. However, one shall note that thrusters are a more expensive controller, because they use fuel that is stored in spacecraft, while the reaction wheel can be powered with electric energy obtained from solar panels.

In order to test the validity of the use of this linear technique for controlling the spacecraft attitude, figure (4) shows a simulation performed with step inputs of amplitude $[\phi_r, \theta_r, \psi_r] = [40°,40°,40°]$. The performance of booth actuators is clearly degraded. Reaction wheels cannot generate a steady response in the observed time horizon. On the other hand, thrusters also generate a very oscillatory behaviour. This test suggests the necessity of use nonlinear control laws during the execution of large angle manoeuvres.

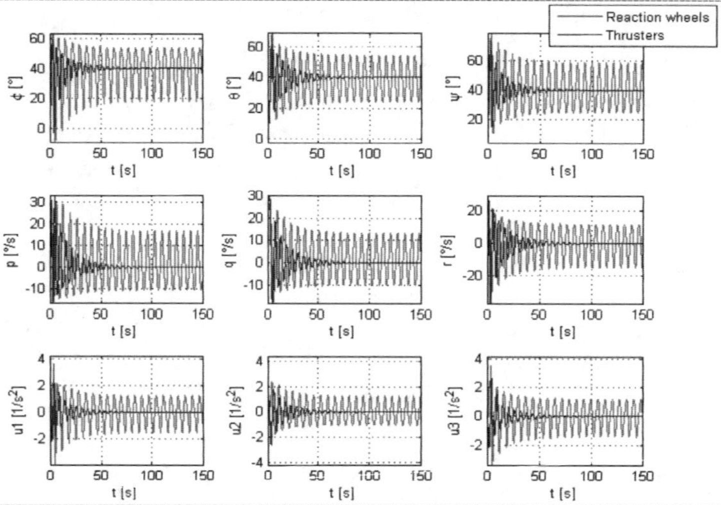

Figure 4. Closed loop spacecraft response to step inputs of high amplitudes.

6. Conclusions

In this work, an attitude controller for the spacecraft in ASTER mission was successfully developed using both reaction wheels and thrusters by means of two distinct loops for stabilization and command tracking with the controller based on Euler angles. A control methodology was used to generate the gains based on a linear quadratic regulator approach with output feedback. The linear model is given by a circular orbit around Alpha ignoring the perturbations due to the gravitational influence of the two remaining satellites: Beta and Gamma. The evaluation was performed via nonlinear simulations, the gravity gradient torque of all the three bodies in the asteroid system was considered. Under small amplitude commands, booth actuators, reaction wheels or thrusters, can provide similar performance. Some differences in settling time are negligible when compared with the orbital period. However, under the application of large amplitude commands, the performance of booth actuators is depreciated, which suggest that nonlinear control laws shall be used for large angle manoeuvres.

References

[1] Sarli B, Winter O, Vieira Neto E, Paglione P 2012 *Strategies for Exploring the Triple System 2001sn263 - Target of the ASTER Mission*(39o COSPAR Scientific Assembly)

[2] Nolan M et al 2008 *Arecibo radar imaging of 2001 SN263: a near-Earth triple asteroid system*(Asteroids, Comets, Meteors) no. 8258

[3] Fang J, Margot J, Brozovic M, Nolan M, Benner L, Taylor P 2010 *Orbits of Near-Earth Asteroid Triples 2001 SN263 AND 1994 CC: properties, origin, and evolution*(The Astronomical Journal) v.141, n. 5, p.141-154

[4] Stevens B, Lewis F 2003*Aircraft Control and Simulation*(John Wiley & Sons, Inc.), chapter 5

[5] Da Silva A, Paglione P, Yoneyama 2012 *Cruising Autopilot for a Flexible Aircraft with Internal Loop of Model Following* (J. Aerosp. Eng.)

Journal of Physics: Conference Series **465** (2013) 012005 doi:10.1088/1742-6596/465/1/012005

Rendezvous maneuvers using Genetic Algorithm

Denílson Paulo Souza dos Santos[1], **Anderson Rodrigo Barretto Teodoro**[2], **Antônio F. Bertachini de Almeida Prado**[1],

[1] INPE - National Institute for Space Research, Mechanics and Control Division, Avenida dos Astronautas, 1758, São José dos Campos.

[2] ETEP / CETEC - Science and Technology Center, Computer Engineering, Avenida Barão do Rio Branco, 882, São José dos Campos.

E-mail: denilson.paulo@gmail.com; andersonrbteodoro@gmail.com; prado@dem.inpe.br

Abstract. The present paper has the goal of studying orbital maneuvers of Rendezvous, that is an orbital transfer where a spacecraft has to change its orbit to meet with another spacecraft that is travelling in another orbit. This transfer will be accomplished by using a multi-impulsive control. A genetic algorithm is used to find the transfers that have minimum fuel consumption.

1. Introduction

To study space maneuvers that search for the minimum fuel consumption in interplanetary missions is an important field of research for the development of space technologies. The problem of transferring a spacecraft between two coplanar circular orbits with free time in a central force field was studied in the present research looking for multi-impulsive transfer using a genetic algorithm. It is obtained as a solution a four-impulsive elliptical transfer orbit [1 – 4]. In this particular case the transfer of the spacecraft is made from one body back to the same body [1]. The literature is extensive with respect to the problems involving transfer orbits and optimal spacecraft maneuvers [1 - 12].

2. Genetic Algorithm

The genetic algorithm is a stochastic global search method inspired on the natural genetic and biological evolution. The genetic algorithm operates on a population of potential solutions by applying the principle of survival of the fitness to produce better and better approximations to find the best solution.

At each generation, it is created new individuals by the process of selecting individuals according to their level of fitness in the problem domain and breeding them together using operators inspired in natural genetics. So, the idea is to create individuals that are better suited to their environment than the individuals that existed in the previous generation.

In this case, a generation is one iteration with a certain quantity of evaluation of the objective function. In each evaluation, the genetic operators are applied, inspired in natural genetics. For the representation, the individuals are created as a chromosome where each one is considered a possible solution for the problem.

Broadly speaking, genetic algorithms differ from traditional techniques by using probabilistic rules, not deterministic, and by working with population of individuals, different of other methods that work from a single point. This is a great advantage, because by maintaining a population with good individuals, the probability of reaching a false peak is reduced.

This algorithm introduces many new concepts, including the fitness of a chromosome that represents the quality of the individual at the environment, the selection probability of an individual, the crossover operator, the mutation and the epidemic operators to introduce random perturbations in the search. In Figure 1 it is demonstrated an analogy to help the understanding of the link between genetic and computer languages.

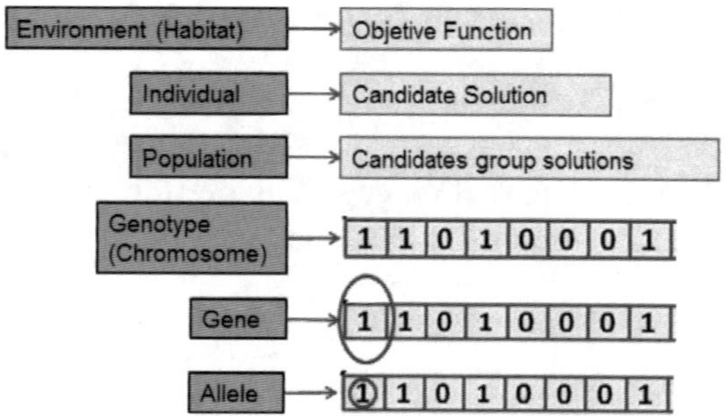

Figure 1. Demonstration of a crossing-over operator between parents.

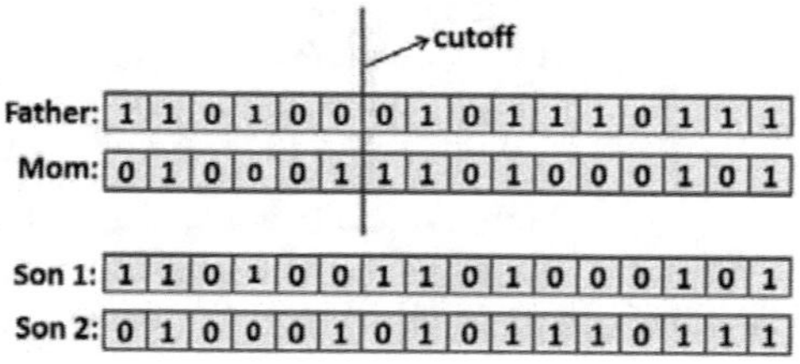

Figure 2. Demonstration of a crossing-over operator between parents.

2.1. *Crossing-Over*

The crossing-over operator is used to create new individuals from parents. This new individual is created from a random cut-off dividing the parents in two parts, having the same or different sizes. The first son consists on the left part of the father with the right part of the mother and the second son consists on the right part of the father with the left part of the mother. Figure 2 demonstrates this schema.

2.2. *Mutation*

The mutation operator is used to introduce random perturbations in the search, having a low occurrence probability, around 0.5%. This perturbation is created by generating a random number between 0 and 1. If this number is smaller than the occurrence probability, the mutation operator works, flipping randomly a bit in the chromosome, if not, the mutation operator is ignored.

2.3. *Epidemic*

The epidemic is another operator used to introduce random in the search, having a very low occurrence probability, around 0.01%. This perturbations is created by generating a random number between 0 and 1, if this number is smaller than the occurrence probability, the epidemic operator works, killing a random part of the population and creating new individuals to occupy the environment.

2.4. *Objective Function*

Comparisons are made for the fitness of the solutions to decide which individual should be propagated to the next generation. Normally, fitness is directly related to the value of the objective solution, with a better objective value indicating the higher fitness. When the Genetics Algorithm (GA) procedure calls its function objective, it passes an array in the first parameter that specifies the selected solution, which is referred as the selection parameter. The selection parameter must not be altered in any way by the function objective.

Journal of Physics: Conference Series **465** (2013) 012005 doi:10.1088/1742-6596/465/1/012005

The fitness of each individual should be computed using the five data that define the problem ($a_1, e_1, a_2, e_2, \Delta\omega$, the first being unit because of the normalization) and the three genes ($v_1, \Delta v, y$) that characterize the individual. One obtains, in sequence [13]:
the true anomaly of the arrival point

$$v_i = v_{i-1} + \Delta v \tag{1}$$

the radii of the departure and arrival point are given by

$$r_1 = \frac{a_1(1-e_1^2)}{1+e_1\cos v_1} \tag{2}$$

$$r_2 = \frac{a_2(1-e_2^2)}{1+e_2\cos v_2} \tag{3}$$

the distance between P_1 and P_2, on each arc, is

$$c = \sqrt{r_1^2 + r_2^2 - 2r_1 r_2 \cos\Delta v} \tag{4}$$

the semi-major axis of the transfer orbit is

$$a_{\min} = \frac{r_1+r_2+c}{4} \tag{5}$$

the distances c_1 and c_2 of P_1 and P_2 from the vacant focus F_2 can be specified by the equations

$$c_i = 2a - r_i \tag{6}$$

The Figure 3 shows a description of several important variables.

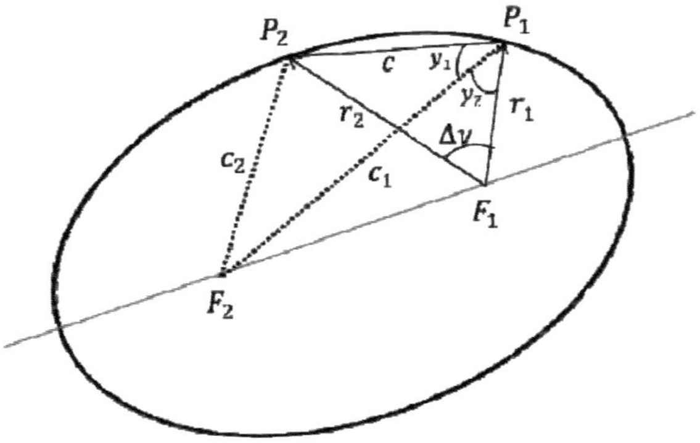

Figure 1 - Geometry of the Problem and the angles involved in the problem.

The angles can be calculated by

$$y = \arccos\left(\frac{r_1^2 - r_2^2 + c^2}{2r_1 c}\right) \tag{7}$$

$$y_1 = \arccos\left(\frac{c_1^2 - c_2^2 + c^2}{2c_1 c}\right) \quad (8)$$

the eccentricity of the transfer orbit is given by

$$e_t = \frac{\sqrt{c_1^2 + r_1^2 - 2c_1 r_1 \cos\gamma_2}}{2a_t} \quad (9)$$

the true anomaly θ_1 of the P_1 on the transfer orbit is

$$\theta_1 = \arccos\left(\frac{a_t(1-e_t^2)-r_1}{r_1 e_t}\right) \quad (10)$$

the argument of perigee for the transfer orbit is

$$\omega = \nu_1 - \theta_1 \quad (11)$$

which is the angle between the perigees of the transfer and the initial orbits.

The geometry of the manoeuvre has been measured. One calculates the radial and the tangential components of the spacecraft velocity before and after both impulses, what permits the computation of the total ΔV, which has been assumed as the measurement of the individual fitness. Non-dimensional variables are used in the code [3]. They are shown below.

$$r = \frac{\tilde{r}}{\tilde{a}_1} \quad (12)$$

$$v = \frac{\tilde{v}}{\sqrt{\frac{\mu}{\tilde{a}_1}}} \quad (13)$$

The distances and velocities references are the semi-major axis of the initial orbit and the velocity of a spacecraft in a circular orbit with the same energy as the initial one. So, the reference time is $\sqrt{\frac{\tilde{a}_1^3}{\mu}}$.

3. Numerical Solutions

Several transfers were simulated with the procedure developed using the genetic algorithm [2 - 4]. The present paper shows one of them, with initial radius $r_o = 2$ and final radius $r_f = 2$. The genetic algorithm provided satisfactory solutions when compared to solutions found in the literature [3 - 5]. The population comprises of 800 individuals and up to 400 generations of individuals (Figure 4).

Journal of Physics: Conference Series **465** (2013) 012005 doi:10.1088/1742-6596/465/1/012005

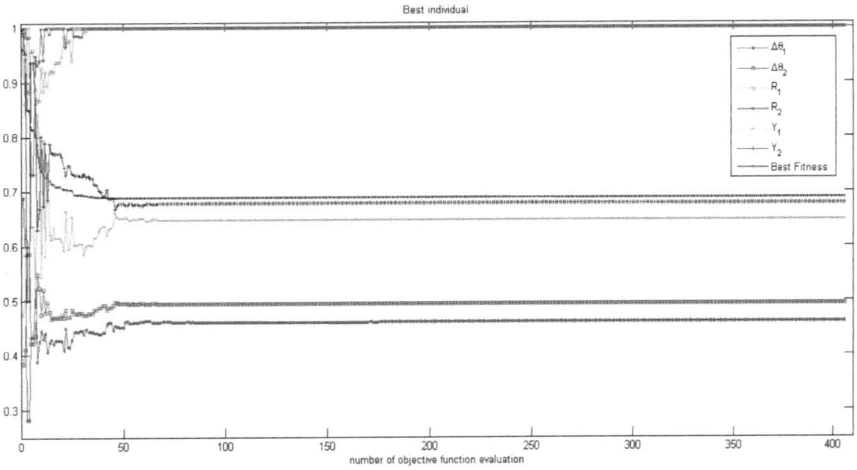

Figure 4 - The variables of the problem using the genetic algorithm method.

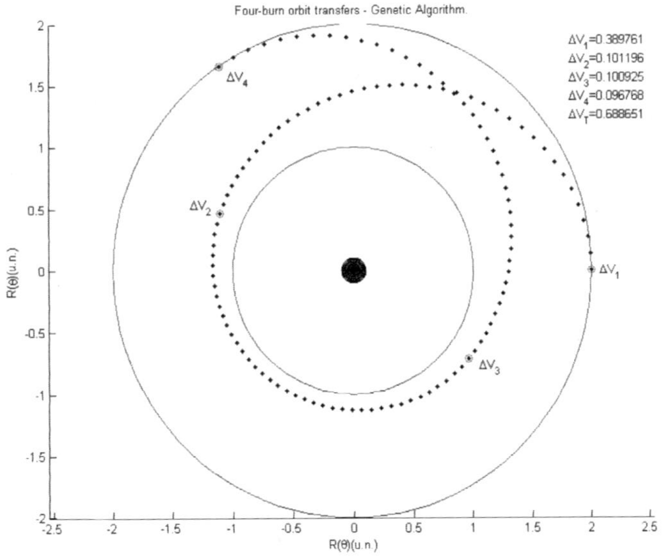

Figure 5 - Transfer Maneuver from one orbit back to the same orbit.

Figure 5 shows a simulation with four-impulses and the total consumption is $\Delta V = 0.688651$ u.n., for a mission leaving and coming back to the same orbit.

4. Conclusion

The rendezvous problem was studied using four-impulsive maneuvers to complete the transfer with the use of genetic algorithm methods. The solutions obtained by the procedure are satisfactory when compared to the classic transfer maneuvers like, for example, Hohmann maneuvers [2].

The results indicate that the maneuvers using the genetic algorithm does not provide better fuel consumption in all cases simulated, but the method proved to find efficient solutions, close to the classical methods for two-impulses, and effective in terms of fuel consumption.

Acknowledgments
This work was accomplished with the support of São Paulo State Science Foundation (FAPESP) under Contract 2009/16517-7 and INPE – National Institute for Space Research – Brazil.

References

[1] Santos D P S, Prado A F B A and Rocco, E. M. 2005 The use of consecutive collision orbits to obtain swing-by maneuvers. *IAF.*
[2] Santos D P S, Prado A F B A and Colasurdo G. 2012 Four-impulsive rendezvous maneuvers for spacecrafts in circular orbits using genetic algorithms. *Math. Prob. Eng.*, 1-16.
[3] Santos D P S and Prado A F B A. 2012 Minimum fuel multi-impulsive orbital maneuvers using genetic algorithms. **AA-AAS-DyCoSS1-11-01**.
[4] Santos D P S, Prado A F B A and Colasurdo G. 2011 Rendezvous maneuvers with minimal ΔV. *Rec. Res. Power Syst. . Syst. Scie., WSEAS.* 38-42.
[5] Prado A F B A. and Broucke R A. 1993 Study of Hénon's orbit transfer problem using the lambert algorithm, *AIAA Jour. Guid. Cont. Dyn.* **17, n5**, 1075-1081.
[6] Gobetz F W and Doll J R, 1969 *AIAA Jour,* **7** 801-834.
[7] Hohmann W, 1925 Die erreichbarkeit der himmelskorper.
[8] Marec, J.P. 1979 Optimal Space Trajectories, (New York: Elsevier).
[9] Santos D P S, Casalino L, Colasurdo G and Prado, A F B A P, 2008 Optimal trajectories using gravity assisted maneuver and solar electric propulsion (sep) towards near-earth-objects, WSEAS *Int. Conf. App. Theor. Mech.* 62-68
[10] Goddard R H, 1919 A method of reaching extreme altitudes. *Smith. Inst. Pub. Misc. Coll.* **71, n.2**.
[11] Chobotov, V.A.; 1996 Orbital Motion – Second Edition. Reston: *American Institute of Aeronautics and Astronautics.*

Journal of Physics: Conference Series **465** (2013) 012006 doi:10.1088/1742-6596/465/1/012006

Re-entry dynamics of high area-to-mass ratio objects in eccentric Earth orbits considering drag and solar radiation pressure

E Bauer Espitia[1] and W Schulz[2]

Orbital Dynamics Group – ODin, Dep. Aeronáutica, FCEFyN, Universidad Nacional de Córdoba – UNC, Argentina

[1] eloybauer@gmail.com
[2] wschulz@efn.uncor.edu

Abstract. The discovery of a new population of objects with high area-to-mass ratios (HAMR) in orbits near the geostationary ring has motivated recent research seeking to understand the behaviour of those debris under the influence of certain types of disturbances, such as solar radiation pressure and atmospheric drag. Based on a series of selection criteria, it was made from the NORAD catalogue a list of potential candidates for study. As we developed an orbital propagator that takes into account the major disturbances, it was possible to study the behaviour and variation of orbital parameters during decay of the HAMR objects. The results obtained for two cases studied exhibit the dynamic behaviour of all orbital parameters and show good agreement with the observed decay times.

1. Introduction

Recent research on the population of artificial objects orbiting the Earth revealed a considerable amount of space debris at high altitudes in the range of dimensions of 10 cm to 1 m [1]. The discovery of these objects with traits in the geostationary transfer area was unexpected in a population where in principle there are no potential progenitors. The orbital periods of these objects are about one revolution per day and their eccentricities are distributed between 0 and 0.6. The most significant characteristic of this new class of debris is that they have area-to-mass ratio so high that are several orders of magnitude larger than those of the "normal" space debris previously cataloged [2]. In turn this supports the hypothesis that the new population corresponds to the debris generated at or near the geostationary ring and are in orbits with varying eccentricities due to disturbances of solar radiation pressure [2].

This work aims to study the dynamics of the high area-to-mass ratio objects (HAMR) not only subject to solar radiation pressure (SRP) as also the air drag in its possible passage through the atmosphere. In this first step, we use the model of cannonball for the SRP and the International Standard Atmosphere to determine the density of the atmosphere at different altitudes into account. In future works the Earth's non-homogeneous gravitational potential will be taken into account and the results will be compared with more accurate models like the ones from Scheeres [3, 4].

Journal of Physics: Conference Series **465** (2013) 012006 doi:10.1088/1742-6596/465/1/012006

2. Equations of motion

Based on an inertial reference system centered on earth we use the following equations to describe the dynamical behavior of HAMR objects [5]:

$$[a]_I = \frac{1}{m_S}[F]_I \tag{1}$$

$$[F]_I = [F]_G + [F]_{SRP} + [F]_D \tag{2}$$

Here $[a]_I$ is the acceleration vector, m_S the spacecraft mass, $[F]_I$ the resultant force vector, $[F]_G$ the gravitational force one, $[F]_{SRP}$ the resultant solar radiation pressure vector and $[F]_D$ the drag force one. See Figure 1.

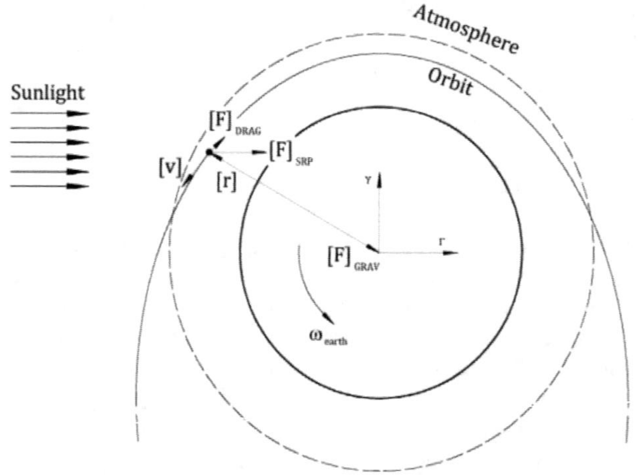

Fig. 1 – Vectorial scheme of forces over HAMR objects.

Additionally:

$$[F]_G = m_S \nabla \frac{GM_\oplus}{r} \sum_{n=0}^{\infty} \sum_{m=0}^{n} \frac{R_\oplus^n}{r^n} P_{nm}(\sin\phi)\{C_{nm}\cos(m\lambda) + S_{nm}\sin(m\lambda)\} \tag{3}$$

For the geopotential we used G for the universal gravitational constant, $M_\oplus$ for the Earth's mass and $R_\oplus$ for its radio. r is the magnitude of the distance between the satellite and the center of the Earth, P_{nm} are the associated Legendre polynomials of degree n and order m, while C_{nm} and S_{nm} stands for the coefficients that describe the dependence on Earth's internal mass distribution. The geocentric latitude is ϕ and the longitude is λ.

$$[F]_{SRP} = \frac{A(1+\rho_S)P_\Phi}{d^2}\hat{d} \tag{4}$$

The solar radiation pressure model considers A the cross-section area normal to the solar flux, ρ_S the reflectivity, P_Φ the solar radiation pressure that is constant and $[d]$ the Sun-satellite vector.

$$[F]_D = -\frac{1}{2}\rho C_D A v_r^2 \hat{\mathbf{e}}_\mathbf{v} \tag{5}$$

While for the drag force ρ is the atmospheric density, the drag coefficient, C_D, is a dimensionless quantity that describes the interaction of the atmosphere with the satellite's material, v_r is the velocity of the vehicle relative to the air and $\hat{\mathbf{e}}_v$ defines the direction of the relative velocity vector.

3. HAMR objects

In order to simulate the dynamical evolution, some special cases were selected based on a set of semi-major axis, eccentricity and area-to-mass ratios according with HAMR debris characteristics. The initial conditions for the tested cases, given in terms of mean Keplerian orbital elements, are provided in Table 1.

NORAD Number	a [km]	e	i [°]	Ω [°]	ω [°]	θ [°]
29343	10899.69	0.38	66.19	29.83	291.47	68.53
29602	9417.10	0.31	39.88	159.33	96.16	263.81

Table 1 – Initial Keplerian orbital elements for the studied HAMR objects.

The initial Keplerian elements were obtained from the corresponding Two Line Elements (TLE), as well as other important parameters for the simulations such as the ballistic coefficient $B*$. The TLE are a mathematical representation of a satellite's mean orbit (https://www.space-track.org/documentation#/tle) for a specific epoch that is frequently updated.

The ballistic coefficient is obtained directly from the object's TLE, anyhow it is necessary to perform a series of manipulations for later use within the code. By definition, we know that $B*$ can be expressed as:

$$B* = \frac{1}{2}\rho_0 C_D \frac{A}{m_S}[R_E^{-1}] \tag{6}$$

Here ρ_0 is the atmospheric density at perigee of the orbit, C_D is the drag coefficient, A/m_S is the area-to-mass ratio of the object and $B*$ is expressed at the TLE in inverted Earth radius R_E units.

For the parameter value A/m_S, necessary in the formulation of the PRS, we used the value of $B*$ and adopted an average value of C_D depending on the height [6]. The area-to-mass ratio found for the studied objects are between 0.1-0.4 m^2/kg.

4. Results

We have simulated the orbital evolution of the two space objects identified by the NORAD numbers 29343 and 29602.

Making use of the initial conditions on Table 1 plus some additional parameters as the specular reflectivity coefficient ($\rho_S = 0.88$) and the aerodynamic drag coefficient ($C_D = 1.8$), propagation was performed for 29343 object from September 4, 2010, 8:32:0.015 UT, the TLE's epoch. The specular reflectivity coefficient and the mean aerodynamic drag coefficient were chosen by the best fit after a series of orbital dynamics simulations on different reasonable values for this kind of object.

The results are in Figure 2, where we can see the variation of the orbital elements until reentry. Secular behavior can be observed with small variations of long period at perigee argument ω, inclination i, and node Ω, while the semimajor axis a and the eccentricity e have a similar behavior until the object enters more deeply inside the atmosphere and these parameters show to be affected more strongly.

In Figure 2 we can see that the ascending node rotates in prograde direction with a rate of change $d\Omega/dt \approx 0.00016506$ [rev/day], while the argument of perigee has a quasi-linear retrograde movement with time variation of $d\omega/dt \approx -0.00097683$ [rev/day], except for the

Journal of Physics: Conference Series **465** (2013) 012006 doi:10.1088/1742-6596/465/1/012006

last five days, where we can see the emergence of a disturbing short component with divergent character. It is further noticed that the orbital inclination decreases very gently with velocity $di/dt \approx -0.000054419$ [rev/day].

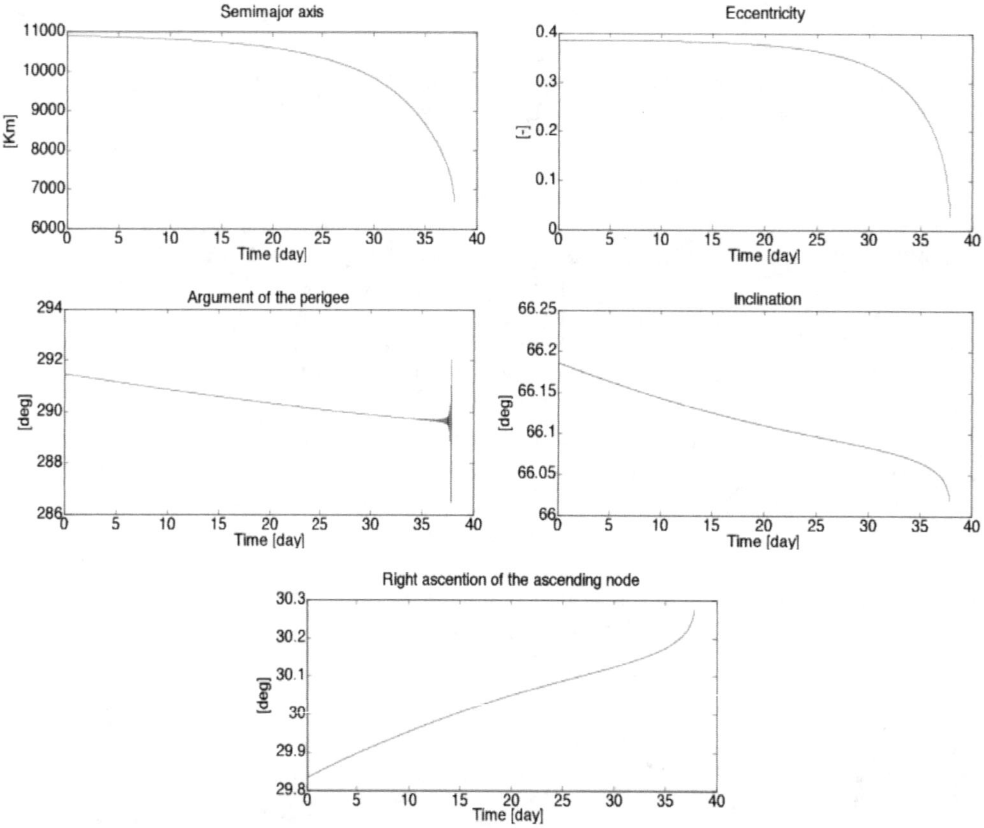

Fig. 2 – Keplerian orbital elements of HAMR 29343 considering SRP and drag effects.

Another significant result can be noticed in Figure 2 and that is the object re-enters after 30.84 days from the initial date, which means that it occurs around October 12, 2010.

For the second object is was made an orbit propagation of the initial parameters according to Table 1 together with the values of specular reflectivity coefficient and drag coefficient average, $\rho_S = 0.21$, $C_D = 1.8$ respectively. Those were chosen with the same kind of simulations performed for the 29343 object. The TLE data epoch used for initialization of the 29602 object trajectory was September 26, 2010 at 23:03:0.495 UT.

The results for 29602 are presented in Figure 3, in which we can see the variation of the orbital elements until re-entry. Secular behavior can be observed with long-period components in the argument of perigee and right ascension of the ascending node. While the semimajor axis, the eccentricity and the inclination, all decrease monotonically, only noticing the atmosphere as moving closer to the Earth's surface and so being strongly affected by the aerodynamic drag force.

Figure 3 shows the ascending node prograde movement of this HAMR object, while the argument of perigee has a smooth variation, except in the last 20 days, when we can see the emergence of a disturbance component with divergent character. It is further noted that the orbital inclination decreases with an approximately constant rate until days before reentry when the rate of decay significantly increases.

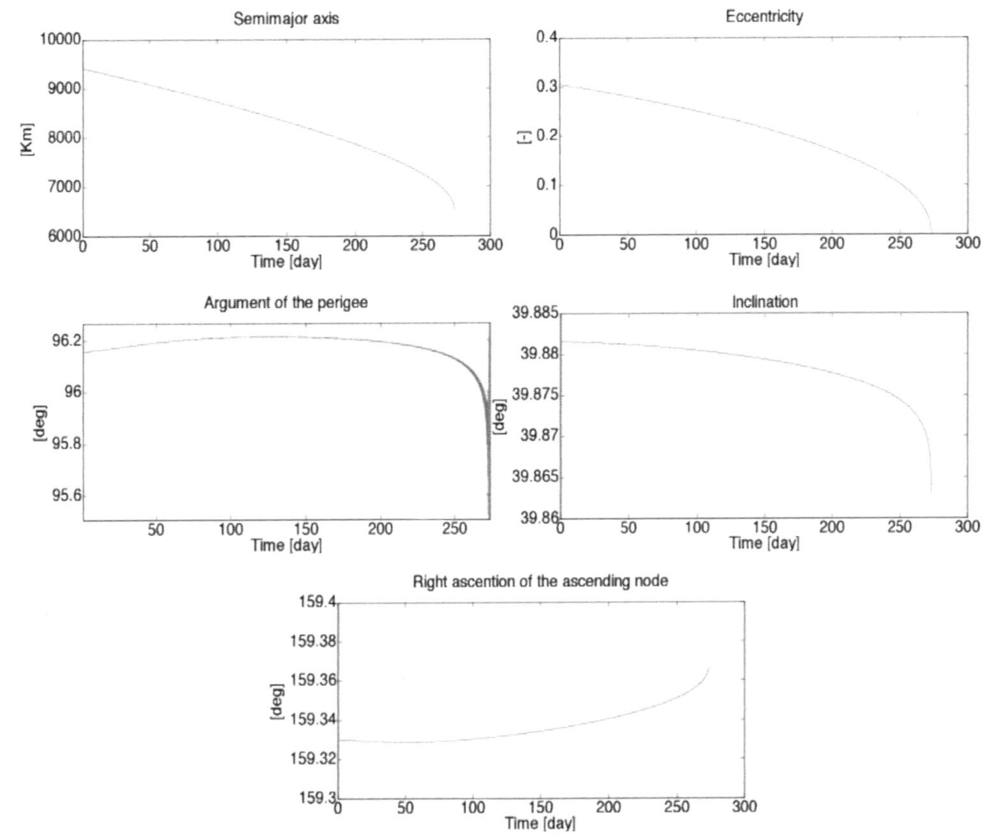

Fig. 3 – Keplerian orbital elements of HAMR 29602 considering SRP and drag effects.

It is also possible to observe in Figure 3 that the object re-enters after 273.2 days from the initial date, which means that it occurs around June 27, 2011. Table 2 summarizes the results relative to the decay time.

NORAD Number	Real decay date	Simulated decay date
29343	October 10, 2010	October 12, 2010
29602	June 21, 2011	June 27, 2011

Table 2 – Real and simulated decay date for the studied HAMR objects.

5. Conclusions

From the results obtained for the two previous cases we may remark that all Keplerian elements behave roughly similarly. As regards the ascending node, its behavior was clearly observed as prograde in both cases with variation rates of the order of 10^{-4} - 10^{-6} [rev/day].

It is possible to notice that the argument of perigee has a smooth retrograde movement in the first case and in the days before reentry, in both cases, a divergent character appears, which can be interpreted as an effect caused by the interaction of the HAMR object with the atmosphere.

As regards the eccentricity, it is noteworthy that the decay observed is the product of the aerobraking effect over the HAMR object that can be interpreted as an orbit circularization. This is a common effect over initially eccentric orbits under the atmospheric drag influence. Furthermore the observed variation in the orbital inclination is caused by the solar radiation pressure experienced during the entire HAMR orbit, because that is the only force that has components outside the orbital plane.

Finally it is important to remark that for the 29343 object the actual re-entry date was October 10, 2010, while for the 29602 object it was June 21, 2011. Comparing the dates obtained by simulated re-entry trajectories with the real ones, as in Table 2, it can be seen that they only differ in 2 days and 6 days respectively, under simulations of around 30 days in the first case, and almost 9 months in the second one. The difference between the decay dates obtained may possibly be due to the simplified models for representing the disturbance forces, as well as not considering the totality of the external forces to which the HAMR objects are subjected. Additionally we should point the fact that we had to estimate and average the values of some important parameters as the drag coefficient and the specular reflectivity coefficient over the trajectories.

For future research we expect to obtain even better results considering the effects of the Earth's non-homogeneous potential. It is also expected that performing re-entry trajectories simulations for a great variety of HAMR objects can lead us to an analytical expression for defining aerodynamic parameters such as the drag coefficient and the specular reflectivity coefficient.

References

[1] Schildknecht T *et al* 2004 Optical observation of space debris in GEO and in hightly-eccentric orbits *Advances in Space Research* **34**

[2] Schildknecht T *et al* 2008 Properties of the high area-to-mass ratio space debris population at high altitudes *Advances in Space Research* **41**

[3] Scheeres D 2007 The dynamical evolution of uniformly rotating asteroids subject to YORP *Icarus* **188**

[4] Scheeres D *et al* 2011 The dynamics of high area-to-mass ratio objects in Earth orbit: the effect of solar radiation pressure *AAS* 11-178

[5] Montenbruck O and Gill E 2001 *Satellite Orbits. Models, Methods and applications* (Berlin: Springer)

[6] King-Hele D 1987 *Satellite Orbits in an Atmosphere: Theory and Applications* (Glasgow: Blakie)

Automatic correction of orbital elements using continuous thrust controlled in closed loop

E M Rocco

National Institute for Space Research-INPE, Space Mechanics and Control Division, São José dos Campos, SP-Brazil

E-mail: evandro@dem.inpe.br

Abstract. This work aims to study and simulate the control of a spacecraft trajectory in order to correct automatically and simultaneously the orbital elements that define the orbit: semi-major axis, eccentricity, periapse argument, inclination and right ascension of the ascending node. Thus, to perform the control of the trajectory was used a propulsion system able to apply thrust with adjustable magnitude and direction of application. In this study it was considered that the propulsion system is controlled in closed loop, so the adjustments of the magnitude and direction of thrust depends on the error generated by comparing a reference state (position and velocity) and a current state. The reference state is determined according to the final orbital parameters. The current state is estimated at each step of the simulation, therefore, the reference and current states must be determined and compared at each step in order to generate the error signal that is inserted into the trajectory control system. However, the control of the orbital parameters simultaneously can be characterized as a multi-objective problem with conflicting goals. The correction of the semi-major axis causes an eccentricity modification and vice-versa. One possibility to deal with this problem is to define when and where to make adjustments for each of the parameters. Thus, the automatic control seeks the best way to correct each parameter, adjusting each one sequentially. At the end of the process all orbital parameters are automatically adjusted and maintained due to the use of the closed loop control system.

1. Simulation and control of the orbital trajectory

The main objective of this study is to simulate the control of the trajectory described by a spacecraft in order to automatically change the orbital elements. The problem can be defined as follows: control the performance of the propulsion system in order to adjusting the orbital parameters that define the orbit.

The trajectory of the vehicle must be controlled through the application of continuous thrust, in a way that at the end of the simulation the final orbit presents the desired orbital elements. To accomplish this task, it was used a control system able to automatically detect the difference between the current state and the desired state of the vehicle, which is a function of keplerian elements of the final orbit. Thus, the simulation occurs discretely, that is, at every simulation step (sampling time) the state of the vehicle is calculated enabling the correction of the orbital velocity.

The orbital motion can be determined solving the Kepler's equation at each step of the simulation. Thus, given an initial state (position and velocity) and the step of the simulation, the state can be converted to keplerian elements using the classical equations from celestial mechanics [1]:

$$M = n(t - T) = u - e \sin u \tag{1}$$

$$n = \sqrt{\frac{\mu}{a^3}} \tag{2}$$

$$r = (X^2 + Y^2 + Z^2)^{1/2} \tag{3}$$

$$v = \left(\dot{X}^2 + \dot{Y}^2 + \dot{Z}^2\right)^{1/2} \tag{4}$$

$$X_i = A_i(\cos u - e) + B_i \sin u \tag{5}$$

$$\dot{X}_i = \frac{a_n}{r}(-A_i \sin u + B_i \cos u) \tag{6}$$

$$\begin{aligned} A_x &= aR_{11}; \quad B_x = a\sqrt{1-e^2}\,R_{12} \\ A_y &= aR_{21}; \quad B_y = a\sqrt{1-e^2}\,R_{22} \\ A_z &= aR_{31}; \quad B_z = a\sqrt{1-e^2}\,R_{32} \end{aligned} \tag{7}$$

$$\boldsymbol{R}(\Omega, i, \omega) = \begin{bmatrix} R_{11} & R_{12} & R_{13} \\ R_{21} & R_{22} & R_{23} \\ R_{31} & R_{32} & R_{33} \end{bmatrix} \tag{8}$$

$$\begin{aligned} R_{11} &= \cos\omega\cos\Omega - \sin\omega\sin\Omega\cos i \\ R_{12} &= -\sin\omega\cos\Omega - \cos\omega\sin\Omega\cos i \\ R_{13} &= \sin\Omega\sin i \\ R_{21} &= \cos\omega\sin\Omega + \sin\omega\cos\Omega\cos i \\ R_{22} &= -\sin\omega\sin\Omega + \cos\omega\cos\Omega\cos i \\ R_{23} &= -\sin i\cos\Omega \\ R_{31} &= \sin\omega\sin i \\ R_{32} &= \cos\omega\sin i \\ R_{33} &= \cos i \end{aligned} \tag{9}$$

$$\vec{h} = \vec{r} \times \vec{v} = \left(Y\dot{Z} - \dot{Y}Z\right)\hat{i} + \left(Z\dot{X} - \dot{Z}X\right)\hat{j} + \left(X\dot{Y} - \dot{X}Y\right)\hat{k} \tag{10}$$

$$r\dot{r} = X\dot{X} + Y\dot{Y} + Z\dot{Z} \tag{11}$$

$$e = \left[\left(\frac{r\dot{r}}{na^2}\right)^2 + \left(1 - \frac{r}{a}\right)^2\right]^{1/2} \tag{12}$$

$$i = \arctan\left[\frac{\left(h_X{}^2 + h_Y{}^2\right)^{1/2}}{h_Z}\right] \tag{13}$$

$$\Omega = \arctan\left(\frac{h_X}{-h_Y}\right) \tag{14}$$

$$f = \arcsin\left(\frac{\left(1 - e^2\right)^{1/2}\sin u}{1 - e\cos u}\right) \tag{15}$$

$$f = \arccos\left(\frac{\cos u - e}{1 - e\cos u}\right) \tag{16}$$

$$\tan\theta = \frac{-\cos i \sin \Omega X + \cos i \cos \Omega Y + \sin i Z}{\cos \Omega X + \sin \Omega Y} \tag{17}$$

$$\theta = \omega + f \tag{18}$$

where M is the mean anomaly; u is the eccentric anomaly; f is the true anomaly; μ is the gravitational constant; a is the semi-major axis; e is the eccentricity; i is the inclination; Ω is the right ascension of ascending node; and ω is the periapse argument.

To simulate and control the trajectory was used the simulation environment STRS (Spacecraft Trajectory Simulator). This simulator is designed to operate in closed loop controlling the trajectory at each instant of time, determined by simulation step defined as one of the input parameters for the simulator [2] [3] [4] [5]. Normally, for correction maneuvers and orbit transfer, it is used a control in open loop commanded by the ground station. However, for maneuvers using continuous thrust, applied during long period of time, the closed loop control of the trajectory becomes advisable, since the errors generated by the non-linearities of actuators and sensors, and further, errors generated by the disturbances could deviate the vehicle from the optimal trajectory. Thus, the control system must be able to detect the deviations from the reference trajectory and produce control signals in order to correct these deviations. A schematic representation of the control system is shown in Figure 1:

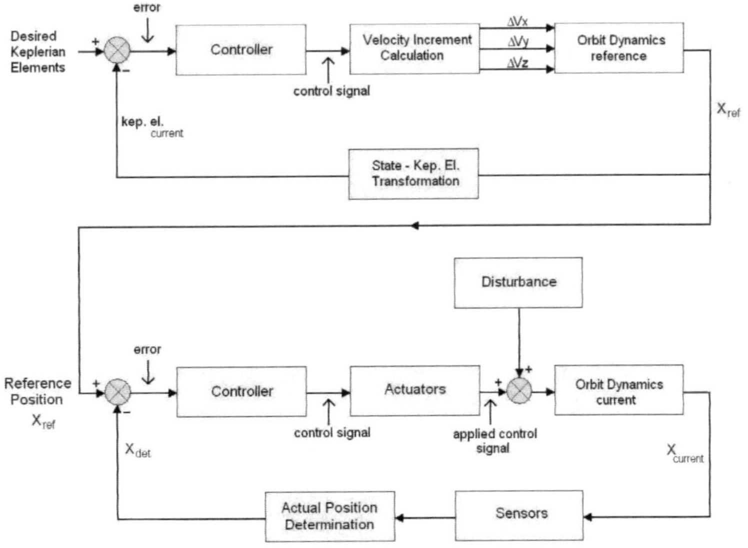

Figure 1. Control system for the trajectory.

In a closed loop control system the reference signal is compared with the signal representing the current state, to generate an error which is inserted into the controller, which generates the control signal. This signal is sent to the actuator to apply it in the dynamic of the system. The output of the dynamic represents the current state that is compared with the reference state, closing the loop.

2. Automatic correction for orbital trajectory

Usually, before the implementation of orbital maneuver, the optimal maneuver is calculated considering the minimization of fuel or time. Generally, obtaining a solution to the problem of optimizing the trajectory is made using numerical methods, which require the determination of the initial values for the variables [6]. However, sometimes, to obtain these initial values is not an easy task. In some cases, the optimal solution cannot be found due to the difficulty of convergence of the numerical methods. When it is desired to correct only one or two orbital elements, a solution can be easily found. But maybe difficult to obtain a solution when it is desired to simultaneously change all elements. Nevertheless, in a real situation the maneuver must be somehow implemented. Therefore, this work presents an alternative to the orbital maneuvers, using sub-optimal maneuvers but in an automatically way, without requiring the use of numerical methods and initial values.

In this work, besides the control loop used to maintain the vehicle close to the best trajectory, was used a control loop to control the variation of each orbital element utilized to obtain the reference trajectory. That is, the final orbital elements are defined, then control loops for each of these elements determine, at each step of the simulation, the magnitudes and directions of application of thrust necessary to reach the desired orbital elements at the end of the maneuver. The variations of elements occur gradually until the difference between the current and reference signals do not generate errors. Then, the propulsion system is turned off automatically.

3. Results

Two simulations are presented. In the first (Table 1 and Figures 2 to 5) the semi-major axis and the eccentricity were simultaneously adjusted. In the second simulation (Table 2 and Figures 6 to 12), beyond the correction of the semi-major axis and eccentricity, the purpose was the changing of the orbital plane by controlling the inclination and the right ascension of the ascending node.

Table 1. Simulation I: correction of the semi-major axis and eccentricity.

initial orbit	**desired final orbit**	**final orbit**
a = 9900000 m	a = 9910000 m	a = 9910000.0016562 m
e = 0.25	e = 0.245	e = 0.24500001
i = 10°		i = 10°
Ω = 55°		Ω = 55°
ω = 25°		ω = 25.80062295°
M = 353.45695°	maneuver time: 10 days	M = 351.84629308°

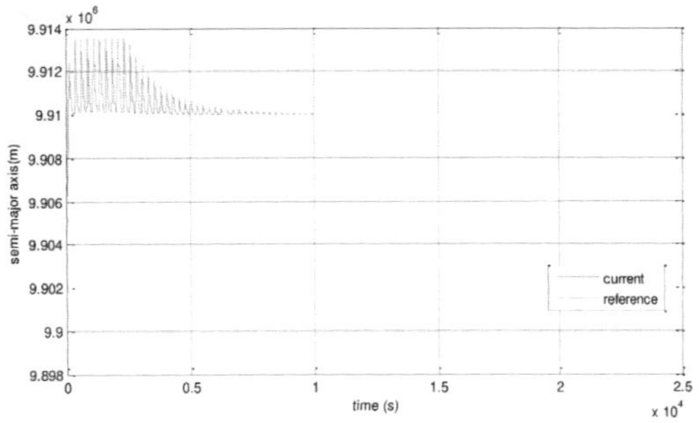

Figure 2. Semi-major axis (I)

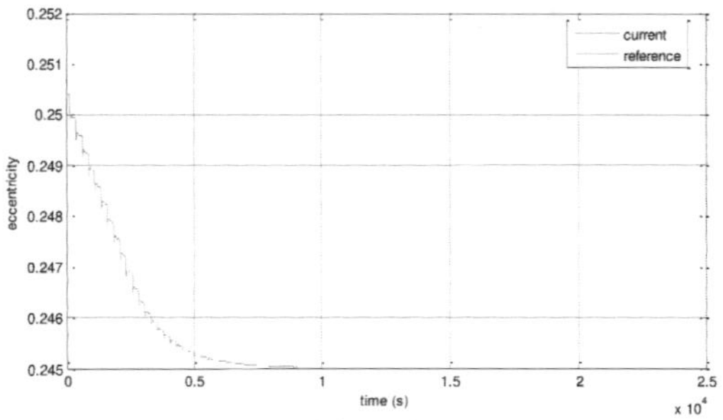

Figure 3. Eccentricity (I)

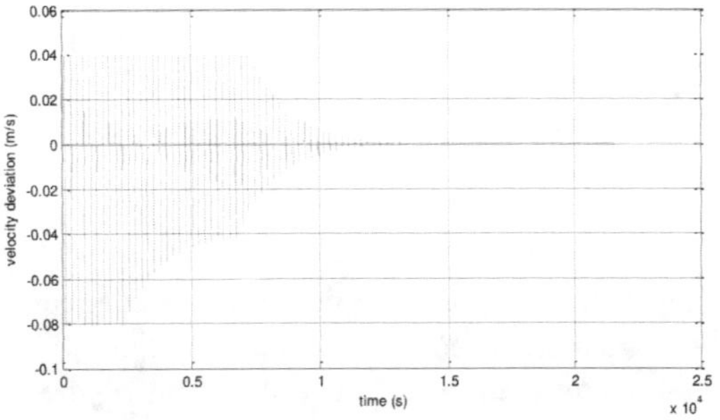

Figure 4. Velocity deviation (I)

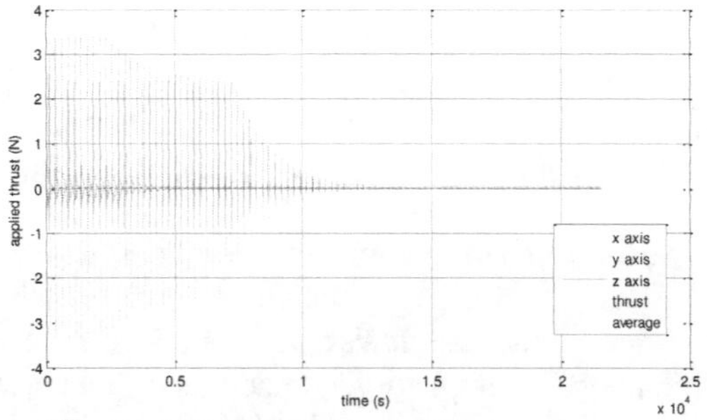

Figure 5. Applied thrust (I)

Journal of Physics: Conference Series **465** (2013) 012007 doi:10.1088/1742-6596/465/1/012007

Table 2. Simulation II: correction of all keplerian elements.

initial orbit	**desired final orbit**	**final orbit**
a = 9900000 m	a = 9910000 m	a = 9910000.4207331 m
e = 0.25	e = 0.245	e = 0.245000212269
i = 10°	i = 10.5°	i = 10.50002165°
Ω = 55°	Ω = 55.5°	Ω = 55.50005465°
ω = 25°	ω = 30°	ω = 29.99963123°
M = 353.45695°	simulation time: 32 days	M = 202.07550398°

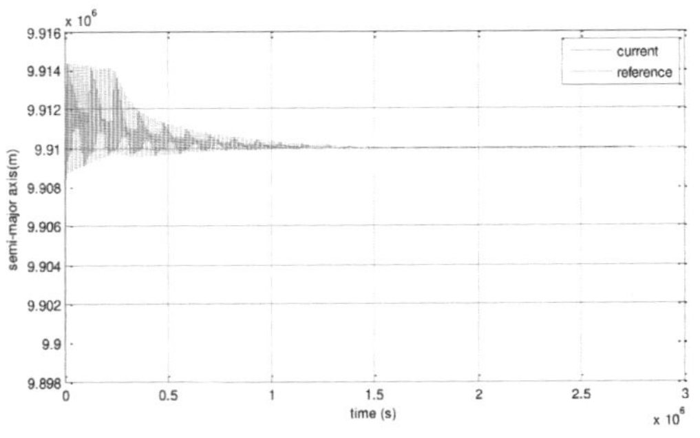

Figure 6. Semi-major axis (II)

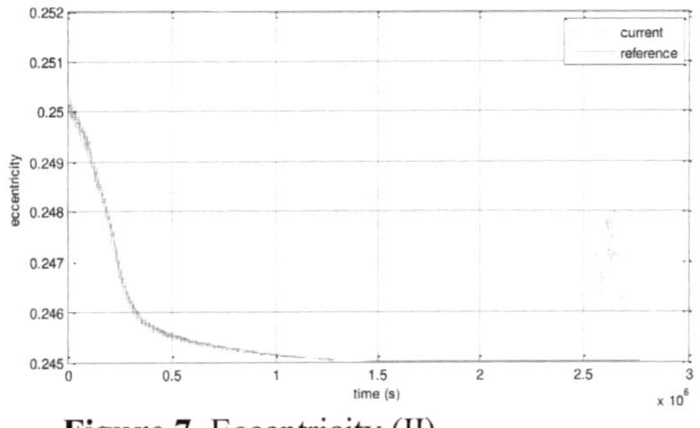

Figure 7. Eccentricity (II)

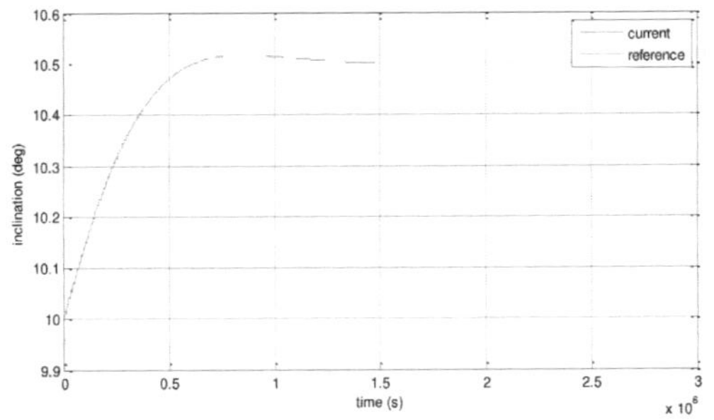

Figure 8. Inclination (II)

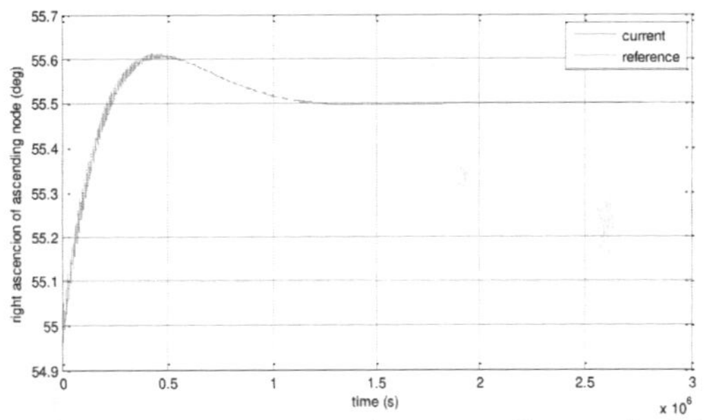

Figure 9. Right ascension ascending node (II)

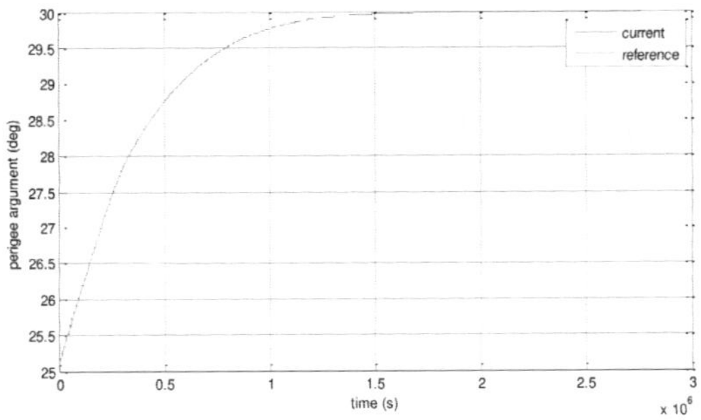

Figure 10. Periapse argument (II)

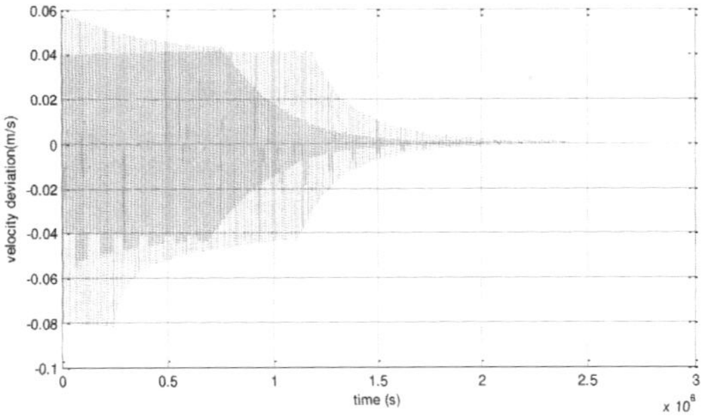

Figure 11. Velocity deviation (II)

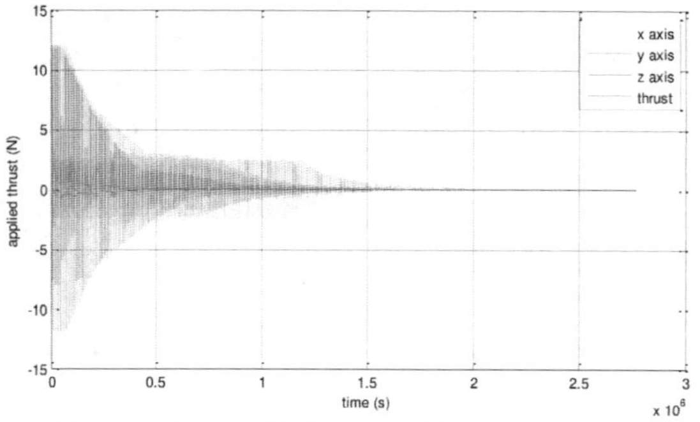

Figure 12. Applied thrust (II)

4. Conclusion

This work represents an alternative to control the trajectory of a spacecraft, especially for long-term missions using continuous propulsion. The closed loop control of the reference allowed the simulating of orbital maneuvers, defined automatically, in order to reach a desired final orbit. Therefore, the automatic procedure to correct the trajectory was implemented and tested successfully. The procedure adopted in this work was able to find sub-optimal maneuvers without the numerical methods and determination of initial values for the solution of these methods.

From simulations was verified that the final orbit was reached with great accuracy, although the time spent in the maneuvers was 10 days for the first simulation and 32 days for the second simulation. However a reduction in the required accuracy could provide a significant reduction of time.

The correction of the orbital elements simultaneously can be characterized as a multi-objective problem with conflicting objectives, because the correction of one element causes a modification on the others. In this work was adopt the automatic control to deal whit the multi-objective problem, so the control system seeks the best way to correct each element, adjusting each one sequentially. A more general study of multi-objective problems can be found in [7].

References

[1] Bate, R.R., Mueller, D.D., White, J.E., (1971). “Fundamentals of Astrodynamics”, Dover Publications, New York.

[2] Rocco, E. M., “Perturbed Orbital Motion with a PID Control System for the Trajectory”, XIV Colóquio Brasileiro de Dinâmica Orbital, Águas de Lindóia, November 17 - 21, 2008.

[3] Rocco, E. M.; Marcelino, E.W.; Prado, A.F.B.A. “Closed Loop Control System Applied to Transfer Maneuvers Using Continuous Thrust”, Mathematical Problems in Engineering, Aerospace and Sciences, São José dos Campos, Brazil, June 30 - July 3, 2010.

[4] Rocco, E. M.; Marcelino, E.W.; Prado, A.F.B.A.; Kuga, H.K. “Closed Loop Control System Applied to Earth-Moon Transfer Maneuvers Using Continuous Thrust”, 38th Scientific Assembly of the Committee on Space Research, Bremen, Germany, July 18-25, 2010.

[5] Rocco, E. M., “Controle de Trajetória com Propulsão Contínua para Missões do tipo Drag-Free”, VII Congresso Nacional de Engenharia Mecânica, São Luís, Brazil, July 31 - August 3, 2012.

[6] Gobetz, F.W.; Doll, J.R., 1969, “A Survey of Impulsive Trajectories”, AIAA Journal, Vol. 7, No. 5: 801-834.

[7] Rocco, E. M., Souza, M. L. O., Prado, A. B. A., “Station Keeping of Constellations Using Multiobjective Strategies”, Mathematical Problems in Engineering, vol. 2013, Article ID 476451, 15 pages, 2013. doi:10.1155/2013/476451

Trajectory control around non-spherical bodies modelled by parallelepipeds

F C F Venditti[1], E M Rocco, A F B A Prado

National Institute for Space Research, INPE, Sao Jose dos Campos, SP-Brazil

E-mail: flaviane.venditti@gmail.com[1]

Abstract. This work is a study of the dynamics around bodies with non-spherical shapes. The gravitational field of a body with non-uniform mass distribution is not central, resulting on orbits around this body that behave different from a keplerian orbit, generating a perturbation on the gravitational field. Irregular objects can be modelled by combining several geometric figures like parallelepipeds. This model can be applied to asteroids, which are objects with non-spherical shapes. The disturbing force generated by these bodies can then be obtained as the sum of the force on each cube. The present paper makes a first study in this problem by showing the evolution of the keplerian elements of orbits around a cube. It also considers a closed loop control system for the trajectory, required to compensate the perturbations due to the irregular shape of the body with the use of a low thrust engine. The main goal is to show that it is possible to use this type of control to keep the orbit keplerian all the time and the details on how to do that.

1. Objectives

The main goal of this work is to study the problem of maneuvering a space vehicle around a body with asymmetric distribution of mass using continuous low thrust in closed loop form for the trajectory.

Some important texts about the formulation for the potential of rectangular objects were developed in the early 90s [1] and continued being developed decades after that [2] [3]. More recently, some progresses on this subject were made [4] [5] [6]. Studies of the property of orbits around elongated rectangular objects were also developed [7].

In this work, the simulator must model the gravitational field of a body following the equations for the potential of a cube, which have a gravitational field that acts different from a spherical body.

Therefore, knowing the deviation on the orbital velocity due to the disturbance of the gravitational field, it is possible to apply the necessary corrections to obtain the desired orbit.

2. Irregular shaped bodies

Asteroids are small bodies orbiting the Sun, with irregular shapes and no atmosphere. Most of the asteroids are located in the Asteroids belt, between the orbits of Mars and Jupiter. A big part of them are carbonaceous, followed by the silicates type, and a smaller part is metallic.

Asteroids with perihelion below 1.3 AU are called Near-Earth Asteroids (NEAs) [8]. Those whose minimum orbit intersection distance with Earth are 0.05 AU, and depending on its absolute magnitude, are considered potentially hazardous asteroids (PHAs). Important data of asteroids were obtained using radar images [9], especially information about the shapes [10].

Journal of Physics: Conference Series **465** (2013) 012008 doi:10.1088/1742-6596/465/1/012008

The gravitational field of an asteroid is not very strong, but when getting closer to the body the perturbation generated by the irregularity of the shape has a stronger effect. For missions planned to analyze closely an asteroid, stay in orbit around a body like this, or even land on it, the study of the gravitational field of the body is essential.

3. Methodology

The gravitational field of the body is modelled as the field generated by a cubic shaped object, assumed to have a constant density. Depending on the distribution of mass and the general shape on an asteroid, it can be represented as the sum of mass concentrations [11][12], or even several geometric shapes, like polyhedrons [13], cubes and other geometric figures that suits best for the case. In this work the cube was chosen for the simulations, using the formulation of its gravitational force previously developed by MacMillan [1]. The reason to choose the cube is that this shape has already several studies in the literature regarding its potential and, since the sizes and the number of cubes used to represent an asteroid can vary arbitrarily, almost all the asteroids can be modelled by this shape. So, as an initial study of this problem, a study regarding the orbital evolution of orbits around one cube is made, as well as a study of the control technique used to keep this orbit keplerian. The technique used here can be applied to asteroids formed by the addition of several cubes.

To solve this problem, several steps need to be made. Given an initial state in Cartesian coordinates, this state can be transformed to keplerian elements and then propagated for the interval of the simulation step. Therefore, the new keplerian elements can be converted again to a new state expressed in Cartesian coordinates.

The algorithm implemented on the simulator should be capable of solving the equations for the transformation of the keplerian elements to the Cartesian state, so it is possible to maneuver the vehicle around a non-spherical body by applying the necessary corrections. The simulator used here is a tool developed by Rocco [14][15][16]. It has several capabilities, like simulating trajectories around spherical and irregular bodies, perform maneuvers of large and small amplitudes, which are correction and transfer maneuvers, besides other applications.

The formulation for the potential, with the keplerian term and a perturbation due to the cubic shape of the object, is given by [1]:

$$U = \frac{Gm}{r} - \frac{7a^4 Gm}{30r^9}[x^4 + y^4 + z^4 - 3(x^2y^2 + x^2z^2 + y^2z^2)] \quad (1)$$

The components of the force (*Fx*, *Fy* and *Fz*) for the perturbation are derived from the formulation for the potential equation (1). They are:

$$Fx = \frac{7a4Gm(x^4 - 5x^2(y^2 + z^2) + (y^4 - y^2z^2 + z^4))}{6(x^2 + y^2 + z^2)^{11\frac{1}{2}}} \quad (2)$$

$$Fy = \frac{7a4Gmy(3x^4 + y^4 - 5y^2z^2 + 3z^4 - x^2(5y^2 + 3z^2))}{6(x^2 + y^2 + z^2)^{11\frac{1}{2}}} \quad (3)$$

$$Fz = \frac{7a4Gmz(3x^4 + 3y^4 - 5y^2z^2 + z^4 - x^2(3y^2 + 5z^2))}{6(x^2 + y^2 + z^2)^{11\frac{1}{2}}} \quad (4)$$

The terms x, y, z are the coordinates for the position of a particle that is travelling around the asteroid, G is the gravitational constant, a and m are the edge and the mass of the cube, respectively.

The total perturbation force (Ft) is given by equation (5):

$$Ft = \sqrt{Fx^2 + Fy^2 + Fz^2} \tag{5}$$

Using the equations of the forces, the components of the velocity change ΔVx, ΔVy, and ΔVz can be obtained. These results make it possible to apply the corrections due to the perturbation. The total variation of velocity to be applied to the satellite (ΔVT) is given by:

$$\Delta VT = \sqrt{\Delta Vx^2 + \Delta Vy^2 + \Delta Vz^2} \tag{6}$$

4. Results

Simulations were made considering the gravitational field of a cube. The initial orbital elements used for the simulations are: semi-major axis of 86 km, eccentricity of 0.13, inclination of 80°, right ascension of the ascending node of 160 °, and argument of periapsis of 72°.

The graphics that will be presented were generated for two different cases:

- *Case I:* trajectories with no control, which shows a comparison of the perturbed orbit and a kleperian orbit of reference. The idea is to show the evolution of the orbital elements and how they differ from a keplerian orbit, when no control is applied to the spacecraft;
- *Case II:* using propulsion, considering the correction of the perturbation of the orbit, so that the vehicle stays in a trajectory close to the keplerian orbit of reference.

Figure 1 shows the orbit around the cube using the initial orbital elements described above. Figures 2 to 11 show the evolution of the orbital elements along the time for case I and case II. Case I considers the orbits showing the gravitational perturbation, and case II using propulsion to correct the effects of the perturbation on the orbit of the spacecraft. The reference orbit is represented in magenta, and the perturbed orbit in blue. The technique used to control the trajectory was to determine the current orbit, considering the perturbation force at each instant of time, and the reference keplerian orbit. Then, by the difference between the current and the reference orbits an error signal was generated. This error signal was used to calculate the necessary thrust force to correct the trajectory in such way that the two forces (perturbation and thrust) have the same magnitude and opposite directions, compensating each other and making the orbit stays close to the keplerian orbit all the time.

Figure 12 shows the magnitude of the thrust applied. Figure 13 is the perturbation graphic, where it is possible to notice that the force of the propulsion is obtained from figure 12, but it is acting contrary to the perturbation, as expected.

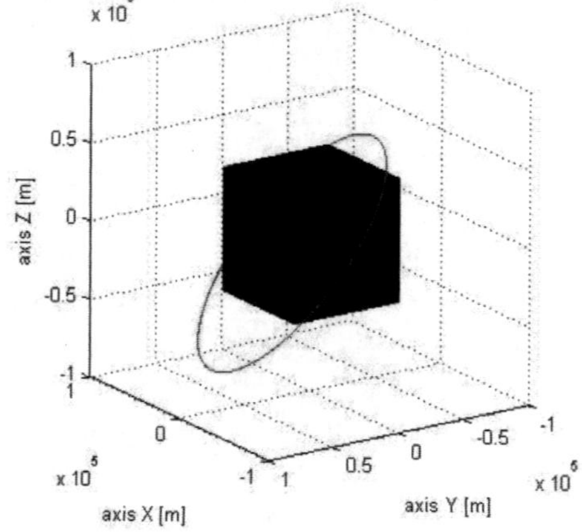

Figure 1. Orbit around a cube.

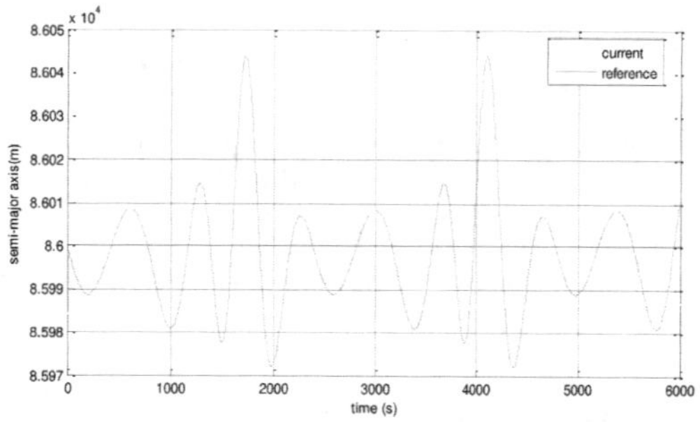

Figure 2. Semi-major axis, case I.

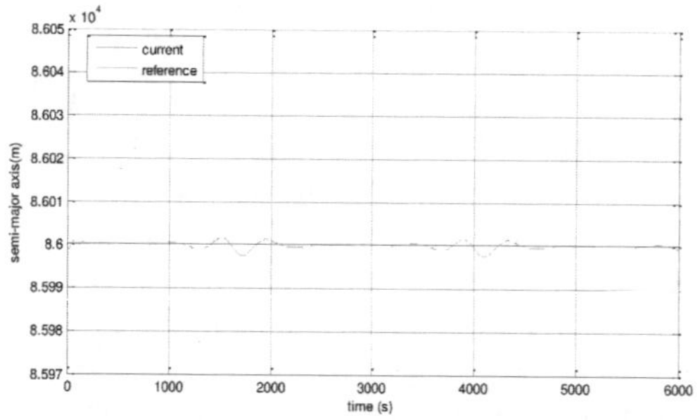

Figure 3. Semi-major axis, case II.

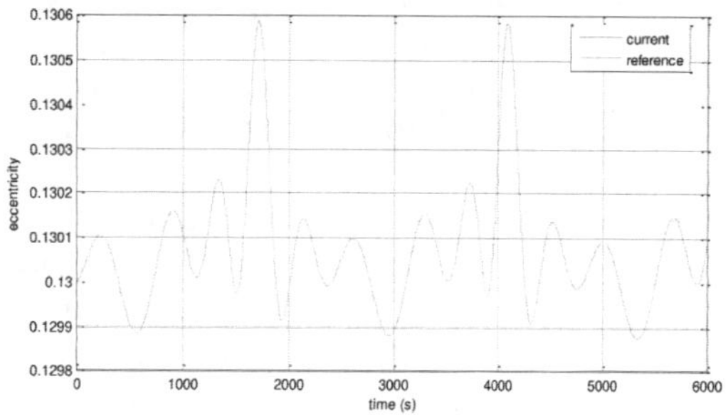

Figure 4. Eccentricity, case I.

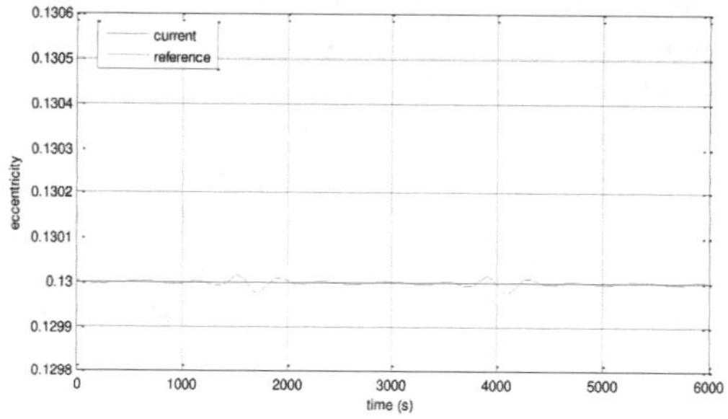

Figure 5. Eccentricity, case II.

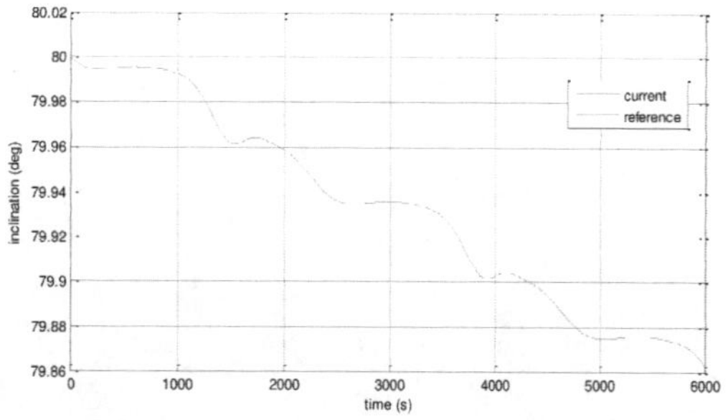

Figure 6. Inclination, case I.

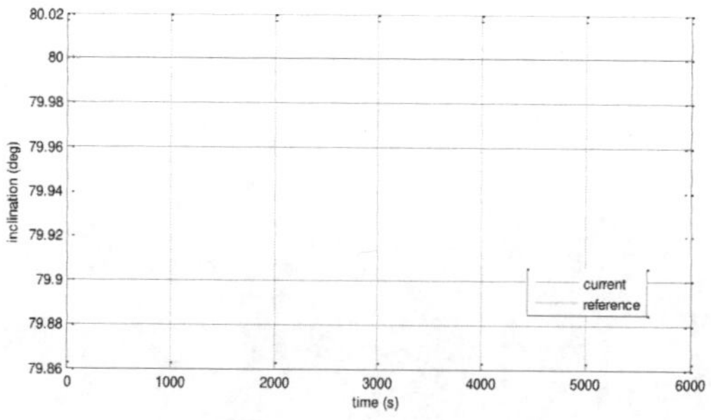

Figure 7. Inclination, case II.

Journal of Physics: Conference Series **465** (2013) 012008 doi:10.1088/1742-6596/465/1/012008

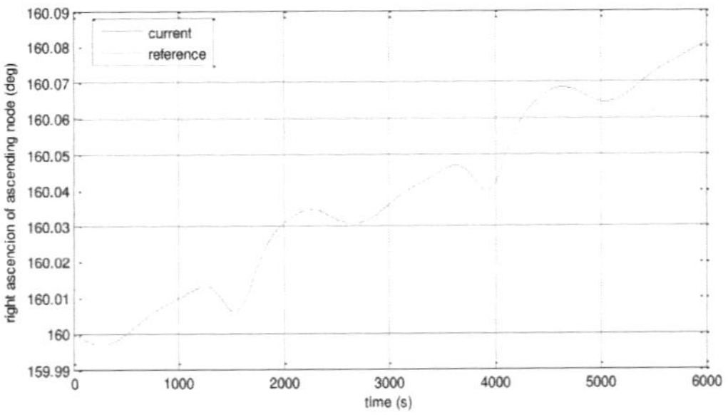

Figure 8. Right ascension of the ascending node, case I.

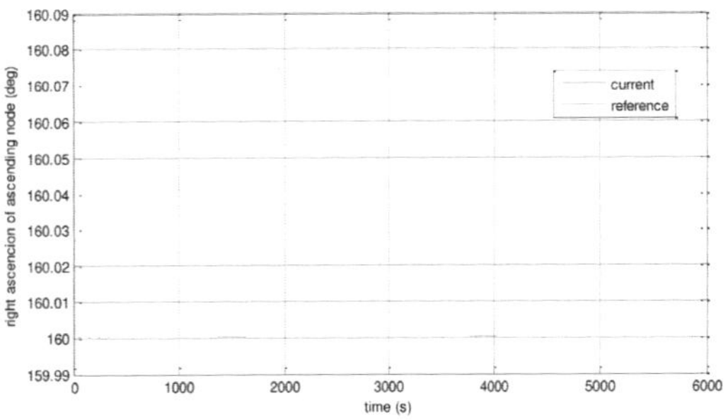

Figure 9. Right ascension of the ascending node, case II.

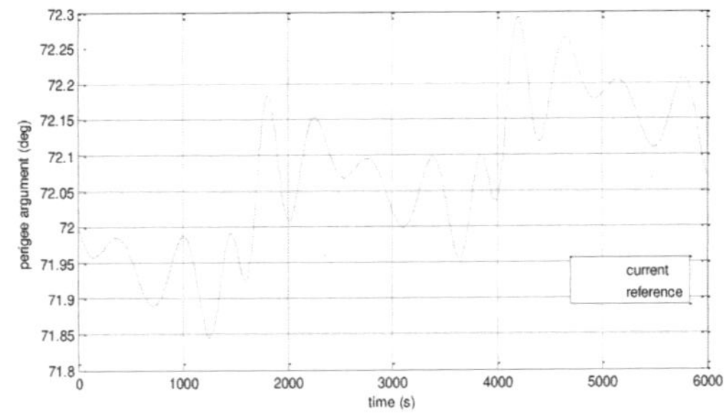

Figure 10. Argument of periapsis, case I.

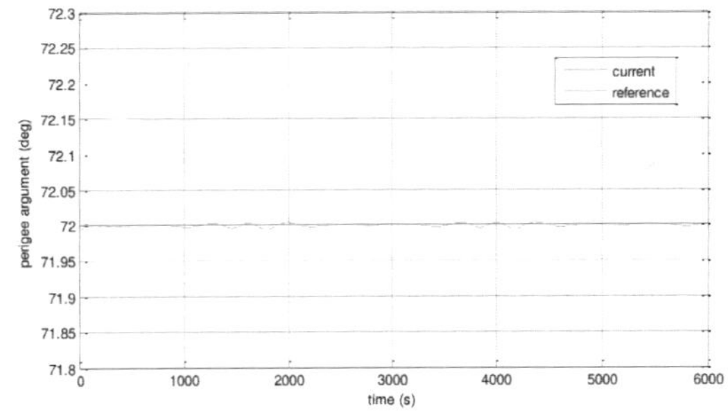

Figure 11. Argument of periapsis, case II.

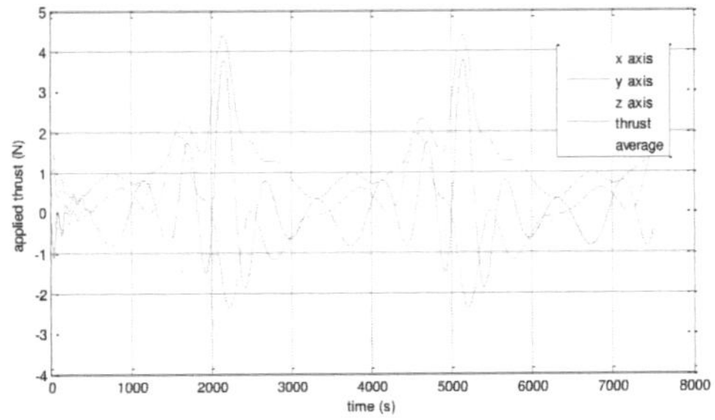

Figure 12. Applied thrust.

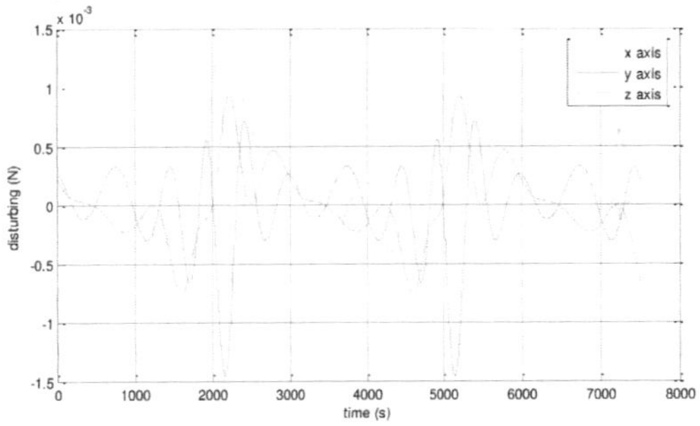

Figure 13. Perturbation.

Case I considers the situation where there is no propulsion applied to correct the perturbation, with the goal of showing the evolution of the orbital elements due to the shape of the object. Case II includes the application of a low thrust in order to control the orbit and to keep the keplerian orbit all the time. The initial inclination and eccentricity chosen, makes the vehicle pass close to the vertices, leading to peaks on the oscillations on the elements during the orbits. Those peaks occur because there are concentrations of extra mass, with respect to a spherical body inside the cube, near the vertices and so the gravitational perturbation near those positions are larger. Considering those peaks, the deviations of the uncontrolled orbit is of the order of 50 km in semi-major axis, 0.0006 in eccentricity, 0.14 degrees in inclination, 0.08 degrees in right ascension of the ascending node and 0.3 degrees in the argument of the periapsis. Simulations using different initial orbits for the spacecraft showed that the effects are stronger, both in terms of the deviations of the uncontrolled orbits or in terms of the fuel consumption to keep the keplerian orbit, when the orbits are closer to the cube.

Journal of Physics: Conference Series **465** (2013) 012008 doi:10.1088/1742-6596/465/1/012008

5. Conclusion

The work presented on this paper is a study of the dynamics of orbits around a non-spherical body, solving the problem of maneuvering a vehicle orbiting this body using continuous low thrust.

To test the capability of the simulator, examples of a cubic shaped object, with edges measuring 80 km, were used. Therefore, the orbit around this body did not behave as a keplerian orbit. Equations for the force generated by a cube were used on the simulations. The lower the orbit, the greater is the perturbation generated by the body.

The examples showed that the simulator is capable of modelling bodies with non-spherical shapes and also control the orbit around the body, proving that the control system is efficient and achieving the goals of the present research. Considerations for small and irregular bodies like asteroids can be created using cubes, and be modelled as the sum of cubes [17], according to the shape of the body.

References

[1] MacMillan W D 1930 *The theory of the Potential* Dover Publications New York 467.

[2] Waldvogel J 1976 The Newtonian Potential of a Homogeneous Cube *ZAMP* **27** 867-871.

[3] Waldvogel J 1979 The Newtonian Potential of a Homogeneous Polyhedra *ZAMP* **30** 388-398.

[4] Werner R A 1994 The Gravitational Potential of a Homogeneous Polyhedron *Cel. Mech.* **59** 253-278.

[5] Werner R A and Scheeres D 1994 Exterior Gravitational of a Polyhedron *Cel. Mech.* **65** 333-344.

[6] Werner R A 1997 Spherical Harmonics Coefficients for the Potential of a Constant Density Polyhedron *Comp. and Geosci.* **23** 1071-1077.

[7] Broucke R A and Prado A F B A 2004 Orbits Around an Elongated 3D-Object *Adv. in the Astronaut. Sci.* **119** 3037-3060.

[8] http://neo.jpl.nasa.gov

[9] Ostro Et al 2004 Radar Observations of Asteroid 25143 Itokawa (1998SF36) *Meteorit. and Planet. Sci.* **39(3)** 407-424.

[10] Scheeres D B, Campbell C, Magri J D Giorgini and D K Yeomans 2000 Radar observations of asteroid 216 Kleopatra *Science* **288** 836-839.

[11] Geissler P, Petit J M, Durda D, Greenberg R, Bottke W, Nolan M and Moore J 1997 Erosion and ejecta reaccretion on 243 Ida and its moon Icarus **120** 140 1997.

[12] Scheeres D J, Marzari F, Tomasella L and Vanzani V 1998 ROSETTA mission: satellite orbits around a cometary nucleus *Plan. and Space Sci. 46*(6), 649-671.

[13] Neese C 2004 Small body radar shape models V2.0. NASA Planetary Data System.

[14] Rocco E M 2006 Tools for Analysis and Simulation of Spacecraft Trajectories in Keplerian Orbit *Technical Report* Center of Applied Space Technology and Microgravity ZARM University of Bremen Germany.

[15] Rocco E M 2008 Perturbed Orbital Motion with a PID Control System for the Trajectory *XIV Colóquio Brasileiro de Dinâmica Orbital* Águas de Lindóia November 17 - 21.

[16] Rocco E M, Marcelino E W, Prado A F B A 2010 Closed Loop Control System Applied to Transfer Maneuvers Using Continuous Thrust *Mathem. Problems in Engin.* Aerospace and Sciences São José dos Campos Brazil June 30 - July 3.

[17] Gaskell R, Saito J, Ishiguro M, Kubota T, Hashimoto T, Hirata N, Abe S, Barnouin-Jha O and Scheeres D 2008 Gaskell Itokawa Shape Model V1.0. *NASA Planetary Data System.*

Journal of Physics: Conference Series **465** (2013) 012009 doi:10.1088/1742-6596/465/1/012009

Analytical Prediction of the Spin Stabilized Satellite's Attitude Using The Solar Radiation Torque

G B Motta[1], M V Carvalho[2] and M C Zanardi[3]
[1,2,3] Group of Orbital Dynamic and Planetology. São Paulo State University, Av. Ariberto Pereira da Cunha, Guaratinguetá, São Paulo, BR

[1]gabriel_borderes@yahoo.com.br
[2]marlondvc@hotmail.com
[3]cecilia@feg.unesp.br

Abstract. The aim of this paper is to present an analytical solution for the spin motion equations of spin-stabilized satellite considering only the influence of *solar radiation torque*. The theory uses a cylindrical satellite on a circular orbit and considers that the satellite is always illuminated. The average components of this torque were determined over an orbital period. These components are substituted in the spin motion equations in order to get an analytical solution for the right ascension and declination of the satellite spin axis. The time evolution for the pointing deviation of the spin axis was also analyzed. These solutions were numerically implemented and compared with real data of the Brazilian Satellite of Data Collection – SCD1 an SCD2. The results show that the theory has consistency and can be applied to predict the spin motion of spin-stabilized artificial satellites.

1. Introduction

The mission accuracy of artificial satellite depends on the position and orientation in a coordinate system fixed in Earth. However the environmental forces and torques acting on the satellite affect the attitude and the orbit of the satellite. The rotational motion is described by the Euler's dynamic equations, which depend on the external torques. Then it is essential to study the influence of the torques in the satellite mission.

The goal of this paper is to analyze the rotational motion dynamics of spin stabilized Earth artificial satellites, through an analytical attitude prediction. Emphasis is given to modeling the solar radiation torque (SRT) and its influence on the spin axis of the spin-stabilized satellite. The direction of the spin axis in relation to the inertial system is specified by the right ascension (α) and the declination (δ).

The Solar Radiation Pressure is created by the continuous photons collisions with the satellite surface, which can be able to absorb or reflect on this flow. The total change of the momentum of all the incident photons on the satellite surface originates from the Solar Radiation Pressure Force, which can cause considerable disturbances in spacecraft orbits when the satellite has a great ratio area/mass. This force can produce a torque.

The most important radiation sources in space are the Sun and Earth. The provided radiation from the Earth has two significant portions: the solar radiation reflected diffusely on the Earth surface (*terrestrial albedo)*, and the *terrestrial radiation*, that is the spontaneous emission in the infra-red range, proportionally to the fourth power of the absolute terrestrial surface temperature. The albedo and the terrestrial radiation are respectively 90 and 93% smaller than the direct solar radiation at 700 km of altitude. Both decrease with the increase of the altitude. So, due to their magnitudes, they are worthless.

In this paper the model of -SRT is given by [1] and their average components for an orbit period are included in the motion equations, therefore it's possible to obtain the analytical solution of these equations.

Two approaches were chosen and used to examine the influence of SRT acting during the evolution of rotational motion of the satellite. In the first approach, the attitude and orbit data were updated every 24 hours. In the second approach, the computed attitude and orbit data weren't updated in order to determine the analytical solution period of validity. In all numerical simulations the orbital elements were updated taking into account the main influence of the Earth oblateness.

Numerical simulations performed with data of Brazilian Satellites SCD1 and SCD2 show the period in which the analytical solution could be used to the attitude propagation, within the dispersion range of the attitude determination system performance of Brazilian Institute for Space Research - INPE.

2. Solar Radiation Torque

A Direct Solar Radiation Torque model was developed in Zanardi and Vilhena de Moares [1] for the case which the illuminated surfaces of the satellite are a circular flat surface, and a portion of the cylindrical surface. It is given by:

$$\vec{N} = N_x \hat{\imath} + N_y \hat{\jmath} + N_z \hat{k} \tag{1}$$

$$N_x = -\frac{k}{R^4}(\beta_1\gamma_1 - \beta_2\gamma_2)\frac{h}{2}\pi\sigma^2 u_z^*\left(u_y^* \cos(\theta) + u_x^* sin(\theta)\right) \tag{2}$$

$$N_y = -\frac{k}{R^4}(\beta_1\gamma_1 - \beta_2\gamma_2)\frac{h}{2}\pi\sigma^2 u_z^*\left(u_y^* \sin(\theta) - u_x^* cos(\theta)\right) \tag{3}$$

$$N_z = 0 \tag{4}$$

$$u_x^* = (a_s R_x + r' a_1) \tag{5}$$

$$u_y^* = \left(a_s R_y + r' a_2\right) \tag{6}$$

$$u_z^* = (a_s R_z + r' a_3) \tag{7}$$

where k is a solar parameter with assumed value given by $1{,}01\text{x}10^{17} kgm/s$, a_s is Sun-Earth distance and here assume the value $1{,}49597870\text{x}10^{11} m$, r' is the satellite geocentric distance, R is the Sun-satellite distance, $\beta_i, \gamma_i, i = 1,2$, are specular and has total reflection coefficients, respectively, for each satellite surface (which assume constant values), a_1, a_2 e a_3 are direction cosines which relate the Orbital System and the Satellite System (given in terms of orbital elements, right ascension and declination of the spin axis [2]), R_x, R_y e R_z give the Sun direction in the Satellite System and are obtained by:

$$\mathrm{R_x} = \mathrm{b_1}\cos(\delta_s)\cos(\alpha_s) + \mathrm{b_2}\cos(\delta_s)\sin(\alpha_s) + \mathrm{b_3}\sin(\delta_s) \tag{8}$$

$$\mathrm{R_y} = \mathrm{b_4}\cos(\delta_s)\cos(\alpha_s) + \mathrm{b_5}\cos(\delta_s)\sin(\alpha_s) + \mathrm{b_6}\sin(\delta_s) \tag{9}$$

$$\mathrm{R_z} = \mathrm{b_7}\cos(\delta_s)\cos(\alpha_s) + \mathrm{b_8}\cos(\delta_s)\sin(\alpha_s) + \mathrm{b_9}\sin(\delta_s) \tag{10}$$

being b_1, b_2, …, b_9 the direction cosines which relate the Equatorial System and the Satellite System (given in terms of satellite's rotation angle, right ascension and declination of the spin axis [2]) and α_s, δ_s the Sun's right ascension and declination, respectively.

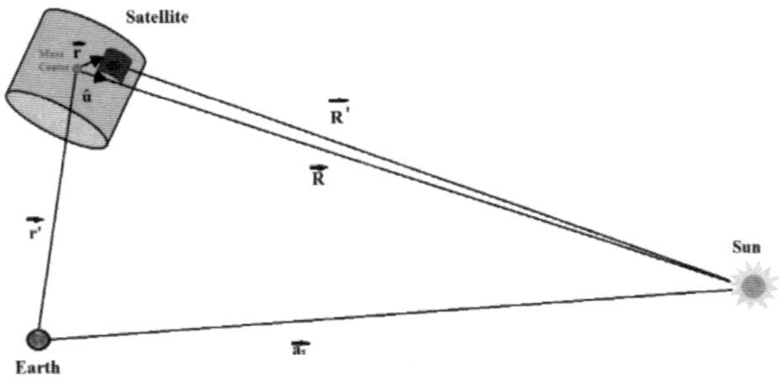

Figure 1. Satellite and Area Element Position

The Sun-satellite distance is represented in Fig. 1 and can be obtained by:

$$R^2 = a_s{}^2 + r'^2 + 2r' a_s \left(a_1 R_x + a_2 R_y + a_3 R_z \right) \tag{11}$$

By the equation (4) it is possible to observe that the component of the satellite axis "Oz" is zero, due to the satellite geometric symmetry (cylindrical shape [1]). Then, the satellite oscillates around this axis.

3. Analytical Solution for the Equations of Rotational Motion

The variations of the spin velocity, the declination and the right ascension of the spin axis for spin stabilized artificial satellites are given by the Euler equations in spherical coordinates [3,4]:

$$\frac{d\alpha}{dt} = \frac{N_x}{I_z \, W\cos(\delta)} \tag{12}$$

$$\frac{d\delta}{dt} = \frac{N_y}{I_z \, W} \tag{13}$$

$$\frac{dW}{dt} = \frac{N_z}{I_z} \tag{14}$$

where I_z is the mean moment of inertia along the spin axis, N_x, N_y, N_z are the components of solar radiation torque.

At first moment it is hard to get the analytical solution for the equations (12) – (14), because the torque components depend on the time. Then in order to solve these equations, in this paper the SRT mean are the components which were used, and obtained by the integration of the instantaneous torque over one orbit period. Now the equations (12) – (14) can be integrated assuming that the orbital elements (I, Ω e ω) were held constant over one orbital period, with the SRT mean components being constant for one orbit. The spin velocity is also constant because the z-torque components are zero:

$$W = W_0 \tag{15}$$

where W_0 is the initial value for spin velocity.

Then the analytical solution for the declination of the spin axis is given by the integration of equation (13) [2]: (16)

$$\delta = \frac{N_{ym}}{I_z W_0} t + \delta_0$$

where δ_0 is the initial value for the declination of spin axis.

After that, by substituting the solution of declination δ in the equation (12), the solution of the right ascension is solved by Motta [2] and given by:

$$\alpha = \frac{N_{xm}}{N_{ym}} ln \left[\frac{sec\left(\frac{N_{ym}}{I_z W_0} t + \delta_0\right) + tan\left(\frac{N_{ym}}{I_z W_0} t + \delta_0\right)}{sec(\delta_0) + tan(\delta_0)} \right] + \alpha_0 \qquad (17)$$

where α_0 is the initial value for right ascension of spin axis.

The solutions presented in the equations (15), (16) and (17), for the spin velocity magnitude, declination and right ascension of the spin axis respectively, are valid for one orbital period. Thus, for every orbital period, the orbital data must be updated, taking into account at least the main influences of the Earth oblateness. With this approach, the analytical theory will be close to the real attitude behaviour of the satellite.

4. Applications

The theory developed has been applied to the spin stabilized Brazilian Satellites SCD1 and SCD2 for verification and comparison of the theory against data generated by the Satellite Control Center (SCC) of INPE. Two approaches were presented [3,5]. In the first one the propagated attitude is daily updated with the help of real satellite data, supplied by INPE. In the second approach daily updates of the attitude data were not performed in the propagation process.

The figures (1) – (4) show the results for the first approach during 40 days. The pointing deviation η is the angle between computed and provided spin axis, and its evolution time is presented in the figures (5) and (6). The results show that the region where the analytical solution is closer to the real data corresponds to the smallest decay of the spin velocity [2]. Over the test period the difference between theory and real data has a mean error deviation in right ascension and in declination of -1,0812° and -0,0092° respectively for the SCD1, and the SCD2 presents a mean error deviation of 0,1666° in right ascension and 0,0584° of mean error deviation in declination. The mean pointing deviation for the period test was 0,3704° for SCD1 and 0,145033° for SCD2 which is within the dispersion range of the attitude determination system performance by SCC (0.5°).

Table 1 presented the results for the second approach. The simulations were interrupted in the 5th day for SCD1 and SCD2 because the mean deviations errors between the computed values and real values for all variables were bigger than INPE's required precision for both satellites.

Table 1. Brazil Satellite's Data.

Day	SCD1's Data			***Day***	SCD2's Data		
	$\Delta\alpha$ (°)	$\Delta\delta$(°)	η (°)		$\Delta\alpha$ (°)	$\Delta\delta$(°)	η (°)
21/08/1993	0	0	0	24/02/2002	0	0	0
22/08/1993	-0,1292	0,2889	0,2898	25/02/2002	0,1744	0,1904	0,2082
23/08/1993	-0,1021	0,5783	0,5786	26/02/2002	0,3937	0,3932	0,4371
24/08/1993	-0,932	0,7084	0,7282	27/02/2002	0,6579	0,6049	0,6842
25/08/1993	-0,4393	0,9781	0,9814	28/02/2002	0,9637	0,8051	0,9321

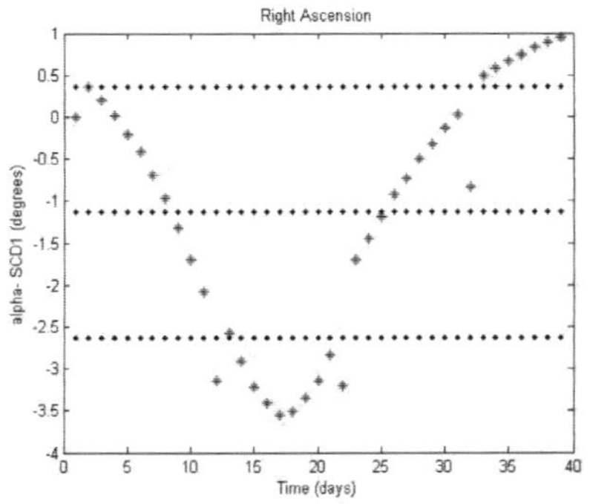

Figure 1. Time evolution for the difference between computed and provided right ascension to SCD1.

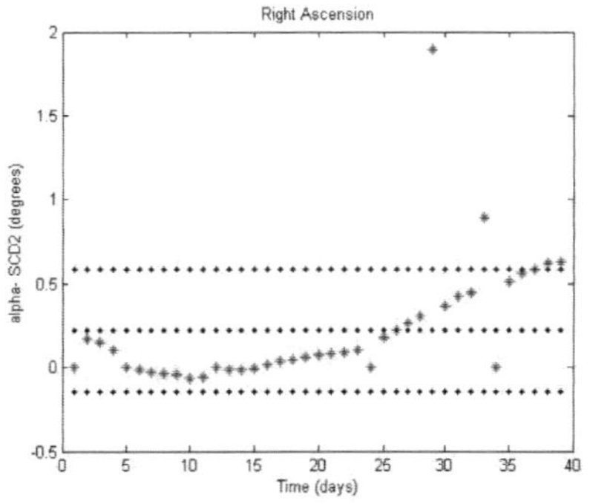

Figure 2. Time evolution for the difference between computed and provided right ascension to SCD2.

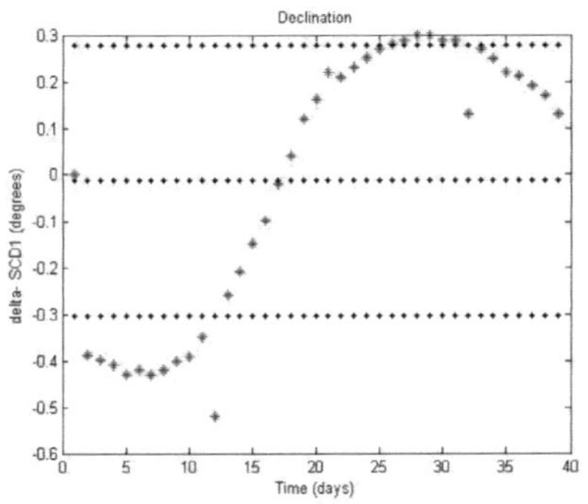

Figure 3. Time evolution for the difference between computed and provided declination to SCD1.

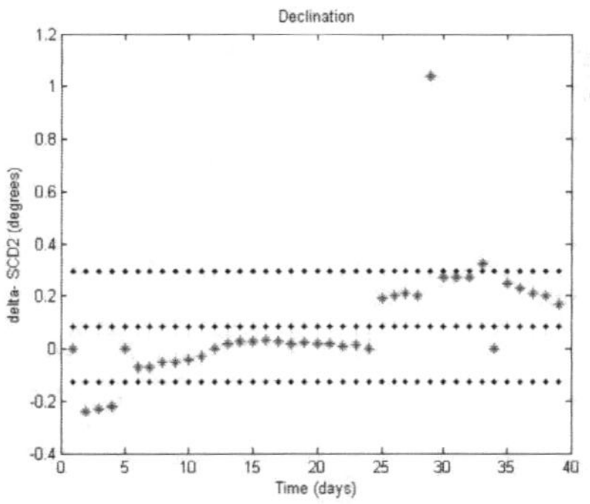

Figure 4. Time evolution for the difference between computed and provided declination to SCD2.

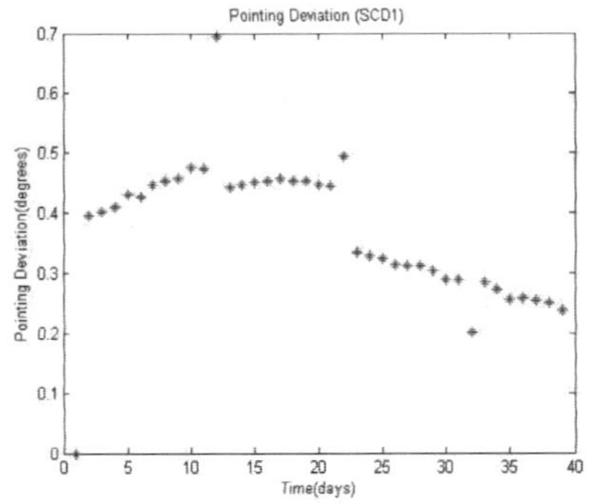

Figure 5. Time evolution for pointing deviation to SCD1.

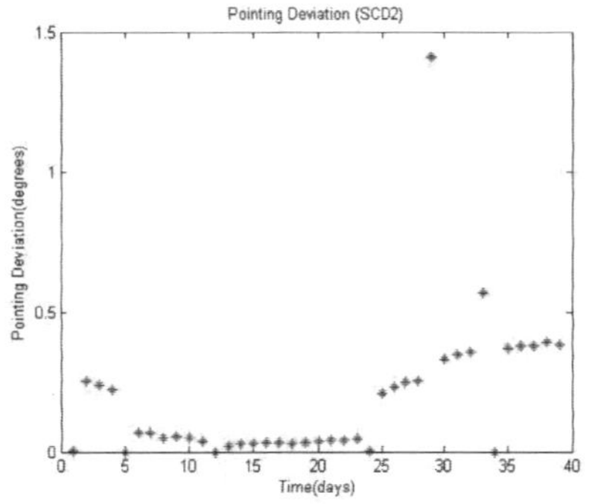

Figure 6. Time evolution pointing deviation to SCD2.

The mean pointing deviation was 0,5156° for SCD1 and 0,4523° for SCD2. The mean declination was 0,5107° for the SCD1 and 0,3987° for the SCD2, which are close to the dispersion range of the attitude determination system performance of SCC, even though the SCD1 surpassed this. The mean right ascension (-0,3205° for SCD1 and 0,4379° for SCD2) is within the required precision by INPE, then it is possible to point out that the declination of the spin axis has great influence in the pointing deviation. The same observation can be applied for the first approach in 40 days for SCD1 simulation where mean pointing deviation and mean declination were within the dispersion range required by INPE while the mean of right ascension did not satisfy the required precision by INPE.

5. Conclusion

In this paper an analytical approach for the spin-stabilized satellite rotational motion was presented taking into account the influence of the solar radiation torque.

The analytical solution shows that the effect of the radiation torque does not cause variation in the spin velocity magnitude because of the satellite's symmetry.

The theory was applied to the spin stabilized Brazilian's satellites SCD1 and SCD2. Results have shown the agreement between the analytical solution and the real satellite behaviour for specific time simulation and two approaches were presented.

In the first one the attitude and orbital data were daily updated with real attitude data supplied by INPE. The results showed a good agreement between the computed and real data during all simulation. The mean pointing deviation was of 0,3704° for the SCD1 and 0,1450° for the SCD2, which are within the dispersion range of the attitude determination system used for this satellite.

In the second approach the attitude and orbital data are not daily updated. The results presented a good agreement between the analytical solution and the actual satellite behaviour only for a five days simulation for SCD1 and SCD2. For more intervals the mean deviation of the right ascension, declination and pointing deviation were higher than the precision required by the SCC/INPE (0.5°).

For both approaches it is possible to note the influence of the declination of the spin axis in the calculation of the pointing deviation. In order to improve the results it is important to include the other external torques. However the procedures are useful for modelling the dynamics of spin stabilized satellite attitude disturbed by solar radiation torque.

6. Acknowledgments

This present paper was supported by CAPES and FAPESP (Proc. n. 2012/21023-6).

7. References

[1] Zanardi M C F P S, Vilhena de Moraes R, *Analytical and Semi-Analytical Propagation of Artificial an Satellite's Rotational Motion,* Celestial Mechanics and Dynamical Astronomy, 75, 227-250, 1999.

[2] Motta G B, Satélites Estabilizados por Rotação e Torque De Radiação Solar Direta. Trabalho de Graduação – Faculdade de Engenharia do Campus de Guaratinguetá, Universidade Estadual Paulista, Guaratinguetá, 2011.

[3] Zanardi M C F P S, PEREIRA, Anderson José . Spin stabilized satellite s attitude analytical propagation.. Proceeding of the 22nd International Symposium on Space Flight Dynamics, 2011, São José dos Campos: INPE, 2011. p. 1-14.

[4] Kuga H K, Silva W C C, Guedes U T V, Dinâmica de atitude para satélites estabilizados por rotação. São José dos Campos – SP, Relatório Técnico do INPE, INPE-4403-NTE/275. INPE – 1987.

[5] Garcia R V, Zanardi M C F P S, Kuga H K . Spin-Stabilized Spacecrafts: Analytical Attitude Propagation Using Magnetic Torques. Mathematical Problems in Engineering (Print), v. 2009, p. 1-19, 2009.

XVI Brazilian Colloquium on Orbital Dynamics IOP Publishing
Journal of Physics: Conference Series **465** (2013) 012010 doi:10.1088/1742-6596/465/1/012010

Analysis of the secular problem for triple star systems

J. P. S. Carvalho,[1] R. Vilhena de Moraes,[1] A. F. B. A. Prado,[2] and O. C. Winter,[3]

[1]UNIFESP-ICT, Universidade Federal de São Paulo, São José dos Campos - SP, Brazil.
[2] Division of Space Mechanics and Control - INPE, São José dos Campos - SP, Brazil
[3] UNESP- Univ Estadual Paulista, Guaratinguetá - SP, Brazil.

E-mail: jeanfeg@gmail.com

Abstract. The long-term dynamics of the three-body problem is studied. The goal is to study the motion of a planet (m_1) around a star (m_0) that is perturbed by a third-body (m_2) (a planet or a brown dwarf star). The gravitational potential is developed in closed form up to the fourth order. Taking into account the triple system, it is shown here the evolution of some orbital parameters of the planet (m_1). A comparison considering models with different orders for the disturbing potential is presented. We show that the behavior of the orbit of the inner planet can flip from prograde to retrograde trajectories. This is due to the third-order term, which strongly affects the eccentricity and inclination. We show that the effect of the fourth order term is to change the times when the phenomenon occurs.

1. Introduction

In the literature, several authors considered the three-body problem to study the dynamics of triple systems, such as [1]-[7]. Here, a triple system is presented as a binary system perturbed by a distant star. In a triple system like this, the distance between the two stars (or planets) is very small when compared to their distances to the third-body (here either a planet or a brown dwarf). Reference [6] presented a study where the authors took into account the octupole term in the disturbing potential. They showed that the inclination of the inner planet varies from prograde ($i < 90°$) to retrograde ($i > 90°$) for a specific problem, where i is the mutual inclination between the two orbits. The authors showed that, considering only the quadrupole term in the potential, the inclination varies according to the Kozai mechanism [8] (see fig. 1 in [6]), i.e., the inclination oscillates with large amplitude when the initial inclination has a value larger than the critical inclination (39.23°) and small amplitudes when the initial inclination has a value smaller than the critical inclination. However, the orbit always remains prograde. When they considered the octupole term in the potential, the inclination grows a lot and can flip from prograde to retrograde trajectories. This mechanism is defined by [9] as eccentric Kozai mechanism (EKM).

In this work we develop the disturbing potential in closed form up to the fourth order in a small parameter ($\alpha = a_1/a_2$), where a_1 is the semi-major axis of the planet and a_2 is the semi-major axis of the disturbing body, to analyze the effects caused by the terms: quadrupole (equivalent to the second order term, the Legendre polynomial P_2), octupole (equivalent to the third order term, the Legendre polynomial P_3) and the P_4 term (order fourth in the Legendre polynomial) on the orbital elements of the planet. The equations of motion are developed in closed form to avoid expansions in power series of the eccentricity and inclination. We compared the models mentioned here in different orders and we also compared with the results obtained by [6]. We investigate what happens in the dynamics when the term of the Legendre polynomial P_4 is considered, and we make various applications to analyze the motion of the planet around a star that is perturbed by a third-body. The approach for the development of the equations is based on [10]-[14]. In order to develop the long-period disturbing potential, the double-averaged method is applied. We analyze the disturbing potential effects on some orbital elements, such as eccentricity, inclination and argument of the periapsis. The subscripts 1 and 2 will refer to the inner and outer orbits, respectively.

2. Equations of motion

The triple system under study is characterized by a body m_1 in an elliptical inner orbit around the body m_0 and a third-body (m_2) moving in an outer elliptical orbit around the center of mass of the system m_0-m_1, but in a very distant trajectory that also has large eccentricity, as shown in Fig. 1. The vector $\mathbf{r_1}$ represents the position of m_1 with respect to the center of mass of the system and the vector $\mathbf{r_2}$ is the position of the body m_2 to the center of mass of the inner orbit. Φ is the angle between $\mathbf{r_1}$ and $\mathbf{r_2}$.

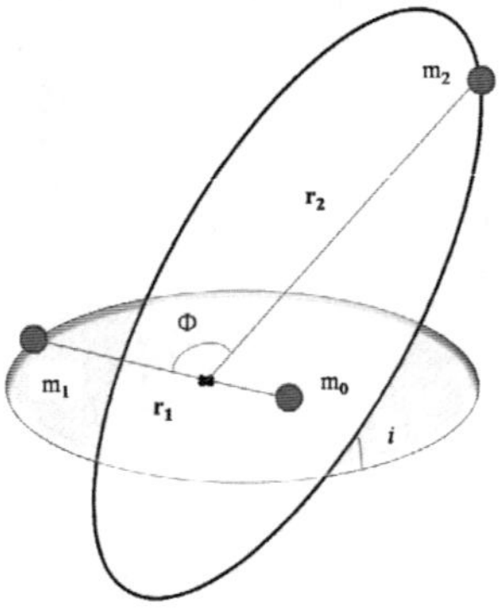

Figure 1. Coordinate system used to describe the triple star system.

The Hamiltonian of the triple system can be written as follows [2, 3, 15]

$$F = \frac{Gm_0m_1}{2a_1} + \frac{G(m_0+m_1)m_2}{2a_2} + \frac{G}{a_2}\sum_{j=2}^{\infty} \alpha^j M_j (\tfrac{r_1}{a_1})^j (\tfrac{a_2}{r_2})^{j+1} P_j(\cos\Phi) \tag{1}$$

where G is the gravitational constant, P_j are the Legendre polynomials and $M_j = m_0m_1m_2\frac{m_0^{j-1}-(-m_1)^{j-1}}{(m_0+m_1)^j}$. We shall deal with the expansion up to the fourth-order in α. The $\cos\Phi$ term is given by [8]

$$\cos\Phi = \cos(f_1+g_1)\cos(f_2+g_2) + c\sin(f_1+g_1)\sin(f_2+g_2) \tag{2}$$

where we will use the shortcut $s = \sin i$, and $c = \cos i$. Here i is the mutual inclination of the inner orbit with respect to the outer orbit. Let g_1, h_1, f_1 and g_2, h_2 and f_2 is the argument of periastron, longitude ascending node and true anomaly of the inner and outer orbits, respectively. The longitude of the ascending node (h) does not appear in Eq (2) because the h_1 and h_2 terms appear as a combination of $h_1 - h_2$. The variables h_1 and h_2 can be eliminated by the theorem of elimination of the nodes [16].

We developed the disturbing potential using a relation given by [9], where g_2 can be written as $g_2 = \pi - h_1$. The disturbing potential is a function of g_1, h_1, so we have the equation of motion to study the behavior of the inclination of the orbit of the planet, as is proved in [9]. For the model considered in the present paper, it is necessary to calculate the terms R_2, R_3 and R_4 of the disturbing function due to P_2, P_3 and P_4 terms, respectively. We get, $R_2 = \frac{G}{a_2}\alpha^2 M_2(\frac{r_1}{a_1})^2(\frac{a_2}{r_2})^3 P_2(\cos\Phi)$, $R_3 = \frac{G}{a_2}\alpha^3 M_3(\frac{r_1}{a_1})^3(\frac{a_2}{r_2})^4 P_3(\cos\Phi)$ and $R_4 = \frac{G}{a_2}\alpha^4 M_4(\frac{r_1}{a_1})^4(\frac{a_2}{r_2})^5 P_4(\cos\Phi)$. The disturbing potential given by Eq. (1) can be written as $F = R_0 + R_2 + R_3 + R_4$, where $R_0 = \frac{Gm_0 m_1}{2a_1} + \frac{G(m_0+m_1)m_2}{2a_2}$. To eliminate the short-period terms of the potential the double-averaged method is applied. Thus, we obtain the disturbing potential expanded up to the fourth order in a small parameter. The long-period disturbing potential (R_2, R_3 and R_4) can be written as

$$R_2 = -\frac{{a_1}^2\mu}{16{a_2}^3(1-{e_2}^2)^{3/2}}[2 + 15\,c^2{e_1}^2\cos(2\,g_1) - 9\,{e_1}^2c^2 - 15\,{e_1}^2\cos(2\,g_1) + 3\,{e_1}^2 - 6\,c^2] \tag{3}$$

$$\begin{aligned} R_3 = \frac{675}{512(1-{e_2}^2)^{\frac{5}{2}}{a_2}^4} e_1\,{a_1}^3\sigma\,e_2 \Big[& \tfrac{7}{9}\,{e_1}^2(c+1)(-1+c)^2\cos(3\,g_1 - h_1) - (c^2 + \tfrac{2}{3}\,c - \tfrac{1}{15})\times \\ & (-1+c)({e_1}^2 + \tfrac{4}{3})\cos(g_1 - h_1) + (c+1)((-\tfrac{7}{9}\,{e_1}^2c^2 + \tfrac{7}{9}\,{e_1}^2)\cos(3\,g_1 + h_1) + \\ & \cos(g_1 + h_1)({e_1}^2 + \tfrac{4}{3})(-\tfrac{2}{3}\,c + c^2 - \tfrac{1}{15}))\Big] \end{aligned} \tag{4}$$

$$\begin{aligned} R_4 = -\frac{\rho\,{a_1}^4}{(1-{e_2}^2)^{\frac{7}{2}}{a_2}^5}\frac{6615}{8192}\Big[& -\tfrac{4}{3}\,(c^2 + c + \tfrac{1}{7})({e_1}^2 + 2)(c-1)^2{e_2}^2{e_1}^2\cos(2\,g_1 - 2\,h_1) - \\ & \tfrac{4}{3}\,({e_1}^2 + 2)(c^2 - c + \tfrac{1}{7}){e_2}^2{e_1}^2(c+1)^2\cos(2\,g_1 + 2\,h_1) + {e_2}^2{e_1}^4(c-1)(c+1)^3\times \\ & \cos(4\,g_1 + 2\,h_1) + {e_2}^2{e_1}^4(c+1)(c-1)^3\cos(4\,g_1 - 2\,h_1) + 4\,(\tfrac{2}{3} + {e_2}^2)({e_1}^2 + 2)(c- \\ & 1)(c^2 - \tfrac{1}{7}){e_1}^2(c+1)\cos(2\,g_1) + \tfrac{10}{7}\,(c-1){e_2}^2(c^2 - \tfrac{1}{7})(\tfrac{8}{15} + \tfrac{8}{3}\,{e_1}^2 + {e_1}^4)(c+1)\cos(2\,h_1) - \\ & 3\,({e_1}^4(c-1)^2(c+1)^2\cos(4\,g_1) + \tfrac{5}{7}\,(c^4 - \tfrac{6}{7}\,c^2 + \tfrac{3}{35})(\tfrac{8}{15} + \tfrac{8}{3}\,{e_1}^2 + {e_1}^4))(\tfrac{2}{3} + {e_2}^2)\Big] \end{aligned} \tag{5}$$

where $\mu = \frac{Gm_0\,m_1\,m_2}{m_0+m_1}$, $\sigma = \frac{Gm_0\,m_1\,m_2\,(m_0-m_1)}{(m_0+m_1)^2}$ and $\rho = \frac{Gm_0\,m_1\,m_2\,({m_0}^3+{m_1}^3)}{(m_0+m_1)^4}$. After some algebraic manipulations we can show that the potential given by Eq. (3) is in agreement with the ones shown in references [4, 9, 13, 17]; the potential given by Eq. (4) is in agreement with the references [9] and [17] and the potential given by Eq. (5) is in agreement with the reference [17]. Therefore, the long-period disturbing potential is written as

$$< F >= R_0 + R_2 + R_3 + R_4 \tag{6}$$

3. Analysis of the long-period disturbing potential

Here we replaced Eq. (6) in the Lagrange planetary equations ([18]) and numerically integrated the set of nonlinear differential equations using the software Maple to analyze the orbital behavior of the body m_1 for some particular cases. We considered the following initial conditions. The star has mass $1M_\odot$, the planet has mass $1M_J$ and the outer brown dwarf has mass $40M_J$. The $M_\odot$ and M_J terms represent the mass of the Sun and Jupiter, respectively. The inner orbit has $a_1 = 6$ AU, $e_1 = 0.001$ and the initial value for the relative inclination is $i = 65°$. These initial conditions were obtained from [6] and are used for all the simulations made here. When we consider the $R_2 + R_3$ terms, the perturbation of the third-body causes the effect of making the inclination of the planet to oscillate between a retrograde and a prograde orbit, as shown in Fig. 2, which confirms the result shown in [6]. Fig. 3 shows the effect when the R_4 term is considered in the disturbing potential. Note that the angular frequency is changed, however, the behavior of the orbit is maintained. In this situation the effects of R_4 is time dependent.

Figs. 4 and 5 show the behavior of the inclination as a function of the time, taking into account the disturbing potential expanded up to the fourth order, and considering different initial values for the argument of the periapsis and for the longitude of the ascending node. It is easy to check the characteristics of the trajectories. As shown in Fig. 4, the inclination oscillates in a prograde orbit for a time longer than the one shown in Fig. 3, and then it starts to oscillate in a retrograde orbit. It also remains in orbit for a time longer than the ones shown in Fig. 2. This difference in terms of orbital behavior is due to the h_1 term that has been changed to another initial condition, here $h_1 = 90°$. Fig. 5 shows that the inclination remains in a prograde orbit for a longer time compared to Figs. 2 and 3, but shorter when compared to Fig. 4.

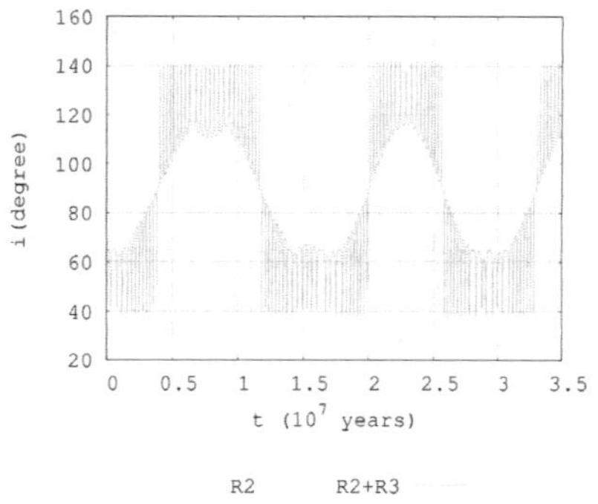

Figure 2. i vs t. Initial conditions: $a_2 = 100$ AU, $e_2 = 0.6$, $g_1 = 45°$ and $h_1 = 43°$.

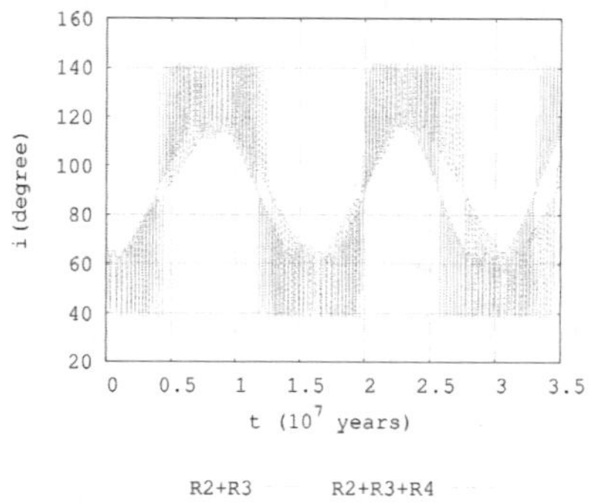

Figure 3. i vs t. Initial conditions: $a_2 = 100$ AU, $e_2 = 0.6$, $g_1 = 45°$ and $h_1 = 43°$.

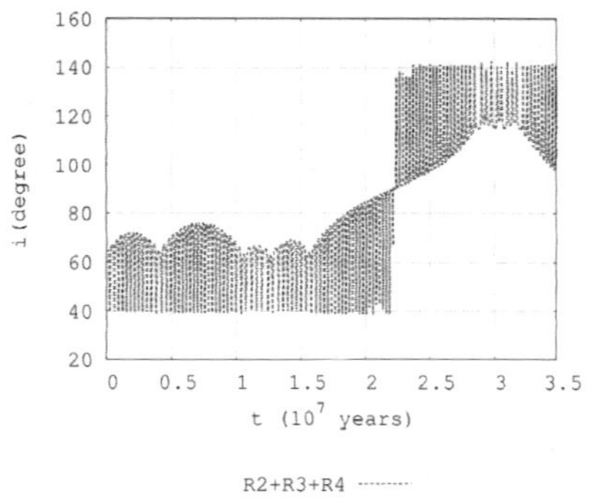

Figure 4. i vs t. Initial conditions: $a_2 = 100$ AU, $e_2 = 0.6$, $g_1 = 45°$ and $h_1 = 90°$.

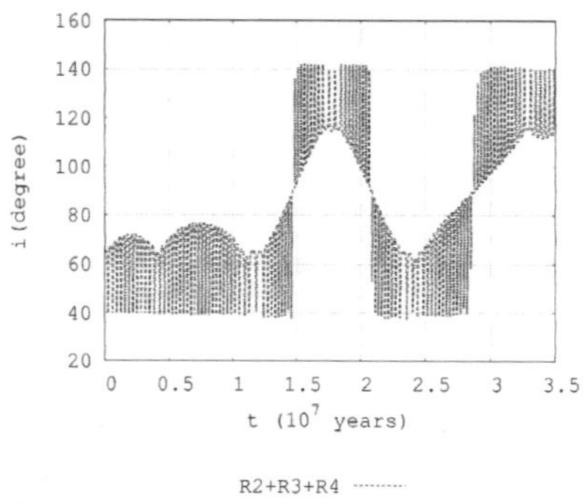

Figure 5. i vs t. Initial conditions: $a_2 = 100$ AU, $e_2 = 0.6$, $g_1 = 65°$ and $h_1 = 90°$.

The diagram e_1 vs g_1 (see Figs. 6 and 7) shows the regions of libration for various values of the eccentricity. For large eccentricities the orbits librates around an equilibrium point for $g_1 = 90°$ with small amplitude, especially for eccentricities $e_1 = 0.6$, 0.7 and 0.8. Fig. 6 were generated taking into account the octupole potential and Fig. 7 were generated taking into account in the potential the R_2, R_3 and R_4 terms. Note that the behavior of the eccentricity in the diagram e_1 vs g_1 (Fig. 7) is a little bit different from the one shown in Fig. 6. The term R_4

Journal of Physics: Conference Series **465** (2013) 012010 doi:10.1088/1742-6596/465/1/012010

has the effect of reducing the amplitude of the librations. These regions increase when g_1 moves away from 90 degrees. This is a consequence of the R_3 and R_4 terms included in the disturbing potential. Note that we found orbits that librate around the equilibrium point for $g_1 = 90°$. This value of g_1 satisfies the condition of frozen orbit ([19]). The search for frozen orbits ([19]) is made looking for inclinations where the eccentricity versus argument of the periapsis diagram has orbits that librate around the equilibrium point. It is interesting to note that Figs. 6 and 7 do not show lines, as Fig. 3 and 4 in [2], but show small regions of libration. This is due to the fact that the inclinations are oscillating between prograde and retrograde orbits.

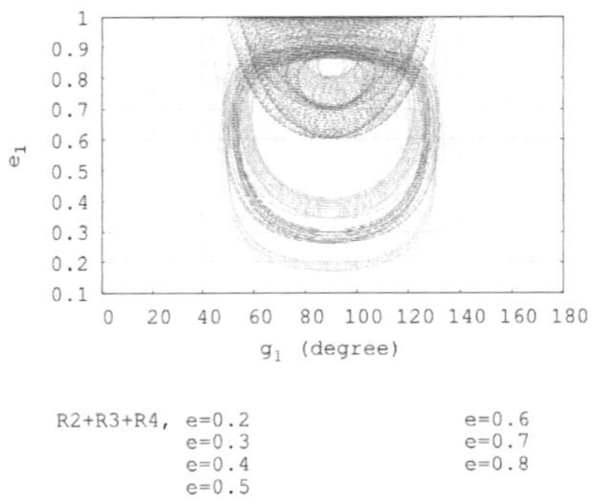

Figure 6. e_1 vs g_1. Initial conditions: $a_2 = 100$ AU, $e_2 = 0.6$, $g_1 = 90°$ and $h_1 = 270°$.

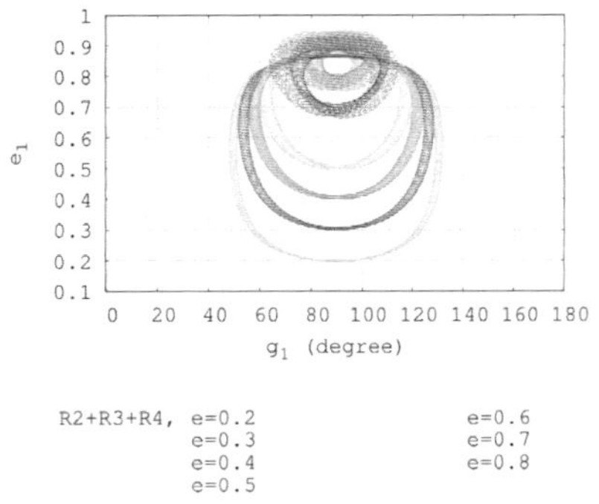

Figure 7. e_1 vs g_1. Initial conditions: $a_2 = 100$ AU, $e_2 = 0.6$, $g_1 = 90°$ and $h_1 = 270°$.

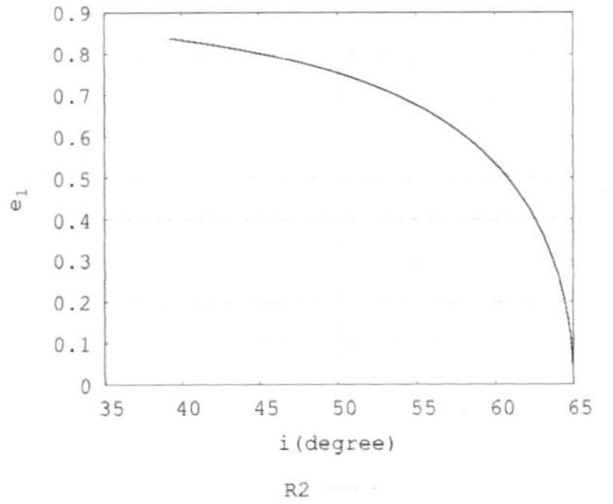

Figure 8. e_1 vs i. Initial conditions: $a_2 = 100$ AU, $e_2 = 0.6$, $g_1 = 45°$ and $h_1 = 43°$.

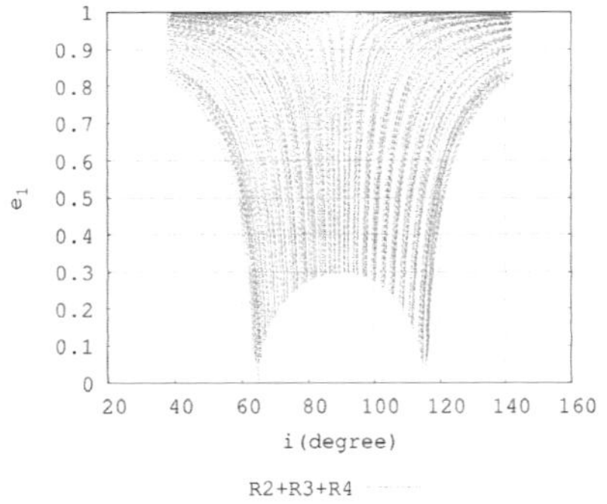

Figure 9. e_1 vs i. Initial conditions: $a_2 = 100$ AU, $e_2 = 0.6$, $g_1 = 45°$ and $h_1 = 43°$.

Fig. 8 shows the behavior of the eccentricity as a function of the inclination. As shown in Fig. 8, when the eccentricity increases the inclination decreases. This is consistent with the Kozai mechanism [8] when taking into account the quadrupole potential. However, including the potential up to the fourth-order in a small parameter (α) (Fig. 9), the inclination shows a different behavior, in comparison with Fig. 8. At the time $t = 0$, the initial mutual inclination is 65°. After a certain time period, the inclination begins to oscillate and the eccentricity reaches a maximum around 0.8, but, when the eccentricity reaches its maximum value the orbit slides to the right, starting to oscillate again. This motion is repeated until near the inclination of 90°. When the inclination reaches this value, the orbit changes its behavior and starts to oscillate in a retrograde orbit. Note also that, in the region near the polar inclination, the eccentricity has a minimum ($e_1 = 0.3$) that is different in comparison with other values of the inclination, as shown in Fig. 9. When the orbit slides right into the polar inclination, the eccentricity reaches a maximum value that is close to unity. At this point, the trajectory approaches an escape orbit and thus the planet may crash onto the star. Because the eccentricity has a large amplitude in its variation, the orbit has a different characteristic when compared to the ones obtained by the Kozai mechanism.

4. CONCLUSIONS

We investigate the dynamics of the three-body problem considering a triple system. We developed the disturbing potential in closed form up to the fourth-order in a small parameter. We analyzed the effects caused by the potential of the third and fourth order, where the dynamics is strongly modified when compared with the Kozai classic problem, where only the potential up to the second-order is considered. We show that the potential developed up to the fourth-order contributes effectively to the inversion of the orbit of the planet from prograde to retrograde. This is due to the strong growth of the eccentricity, inclination and argument of the periapsis caused by the R_3 and R_4 terms.

We show that the behavior of the inclination is sensitive to the initial conditions and that the behavior of the orbit is completely changed when the initial conditions are modified. We show that the effect of the fourth order term preserves the characteristic of oscillating the inclination between a prograde and a retrograde orbit, however, the angular frequency is changed. We found that the R_4 term changes the time when the phenomenon occurs. Taking into account the potential up to the fourth-order, we show a diagram e_1 vs g_1 where our results do not show lines (see Figs. 6 and 7), as Fig. 3 and 4 in [2], but show small regions of libration. This is due to the fact that the inclinations are oscillating between prograde and retrograde orbits. These regions increase when g_1 moves away from 90 degrees.

4.1. Acknowledgments

This work was accomplished with support of FAPESP under the contracts N° 2011/05671-5 and 2012/21023-6 SP-Brazil, CAPES, and CNPQ (300952/2008-2).

5. References

[1] Krymolowski, Y., Mazeh, T., 1999, *Mon. Not. R. Astron. Soc.*, 304
[2] Ford, E. B., Kozinsky, B., Rasio, F. A., 2000, *A P J*, 535
[3] Ford, E. B., Kozinsky, B., Rasio, F. A., 2004, *A P J*, 605
[4] Takeda, G., Kita, R., Rasio, F. A., 2008, *A J*, 683
[5] Correia, A. C. M., Laskar, J., Farago, F., Boué, G., *Celest. Mech. Dyn. Astron.*, 111
[6] Naoz, S., Farr, W. M., Lithwick, Y., Rasio, F. A., Teyssandier, J., 2011, *Nat*, 473
[7] Katz, B., Dong, S., Malhotra, R., 2011, *P R L* **107**, 181101
[8] Kozai, Y., 1962, *A J*, **67**, 9
[9] Lithwick, Y., Naoz, S., 2011, *The Astrophysical Journal* **742** 94
[10] Brouwer, D., 1959, *A J*, **64**, 9
[11] Hori, G., 1961, *A J*, **66**, 6
[12] Prado, A. F. B. A., 2003, *J. Guid. Control. Dynam.*, **26**, 1
[13] Yokoyama, T., Santos, M. T., Gardin, G., Winter, O. C., 2003, *A&A*, 401
[14] Yokoyama, T., Vieira Neto, E., Winter, O. C., Sanchez, D. M., Brasil, P. I. O., 2008, *Math. Probl. Eng.*, 2008, Article ID 251978
[15] Harrington, R., 1969, *Celest. Mech.*, 1
[16] Jefferys, W. H., Moser, J., 1966, *A J*, **71**, 7
[17] Laskar, J., Boué, G., 2010, *A&A*, **522**, A60
[18] Kovalevsky, J., 1967, Introduction to Celestial Mechanics, Bureau des Longitudes, Paris, p. 126
[19] Carvalho, J. P. S., Vilhena de Moraes, R., Prado, A. F. B. A., 2010, *Celest. Mech. Dyn. Astron.*, **108**, 4

Journal of Physics: Conference Series **465** (2013) 012011 doi:10.1088/1742-6596/465/1/012011

Searching sequences of resonant orbits between a spacecraft and Jupiter

J K S Formiga and A F B A Prado
Space Mechanics and Control Division, National Institute for Space Research -INPE, São José dos Campos-SP, BR.

E-mail: jkennety@yahoo.com.br, prado@dem.inpe.br

Abstracts: This research shows a study of the dynamical behavior of a spacecraft that performs a series of close approaches with the planet Jupiter. The main idea is to find a sequence of resonant orbits that allows the spacecraft to stay in the region of the space near the orbit of Jupiter around the Sun gaining energy from each passage by the planet. The dynamical model considers the existence of only two massive bodies in the systems, which are the Sun and Jupiter. They are assumed to be in circular orbits around their center of mass. Analytical equations are used to obtain the values of the parameters required to get this sequence of close approaches. Those equations are useful, because they show which orbits are physically possible when taking into account that the periapsis distances have to be above the surface of the Sun and that the closest approach distances during the passage by Jupiter have to be above its surface.

Key words: Astrodynamics, Orbital Maneuvers, Close Approach Maneuvers, Space Trajectories.

1. Introduction

The description of the close approach is shown in references [1-3]. There are also several other studies that can be mentioned in this topic, like the study of transfer orbits to find trajectories linking the Lagrangian points and the primaries of the system [4-5] or the existence of atmospheric drag during the maneuver [6]. In terms of practical missions that used this technique, it is possible to mention references [7]-[16], which show several applications of the close approach in orbital maneuvers. Some other studies considered variations of this problem, like the combination of impulsive maneuvers during the close approach [17], noncircular orbits for the primaries [18] or the existence of a cloud of particles resulting from an explosion of a main body [19]. For a maneuver that needs to move a spacecraft that is around Jupiter to send it to another location of the Solar System or beyond, it is possible to mention reference [20]. A three dimensional swing-by is described in reference [21]. In 1968, it was introduced in the literature maneuvers with multiple encounters, like in reference [22]. Consecutive meetings between M_3 and the orbit of M_2 in the restricted three-body problem for the case $\mu = 0$, including elliptical orbits for the primaries, is shown in reference [23].

The present paper is a sequence of those studies, because it extends the previous researches to consider the possibility of more than one close approach with a celestial body. The goal is to search for series of close approaches with Jupiter to find trajectories for missions that want to study the space in the Solar System near the planet by changing its orbit, but without escaping to the interstellar space. The orbital elements, energy, angular momentum and velocity of the spacecraft with respect to the Sun are changed by each of these approaches, and the goal is to study their orbital behavior to know how this trajectory will evolve. The dynamical system given by the "patched-conics" is used and the motion is assumed to be planar. Jupiter is assumed to be in a circular orbit in the present model. Extensions to elliptic orbits were considered, but the results do not have large influences. Figure 1 shows a sketch of the maneuver.

So, this technique uses close approach maneuvers to change the orbit of the spacecraft, to allow it to pass by different parts of the space near the orbit of the planet Jupiter around the Sun. It means that, by changing the orbital elements of the spacecraft, its path in the space covers different areas, so it is possible to make measurements of important scientific data like temperature, radiation level, density of particles and others in different areas without using fuel expenditure to change the orbit of the spacecraft. A complete description of the evolution of the orbits, showing which values of semi-major axis, eccentricities, perigee and apogee distances are covered is also made, in order to known exactly the regions of the space that are included in the trajectories found.

2. Mathematical model

This problem can be studied assuming a system formed by three bodies: The Sun (M_1), Jupiter (M_2) and a spacecraft with an infinitesimal mass (M_3) that remains orbiting the Sun and then makes a close approach with Jupiter. The following variables are used to identify one close approach trajectory: r_{ap} (the distance from the spacecraft to the center of Jupiter at the moment of the closest approach), $\vec{v}_\infty^-$ and $\vec{v}_\infty^+$ (velocities of the spacecraft with respect to Jupiter, before and after the maneuver, respectively, in the inertial frame), $\vec{v}_2$ (velocity of Jupiter with respect to the Sun), δ (half of the angle of the curvature due to the close approach) and ψ (angle of approach). Reference [3] shows more details on that problem.

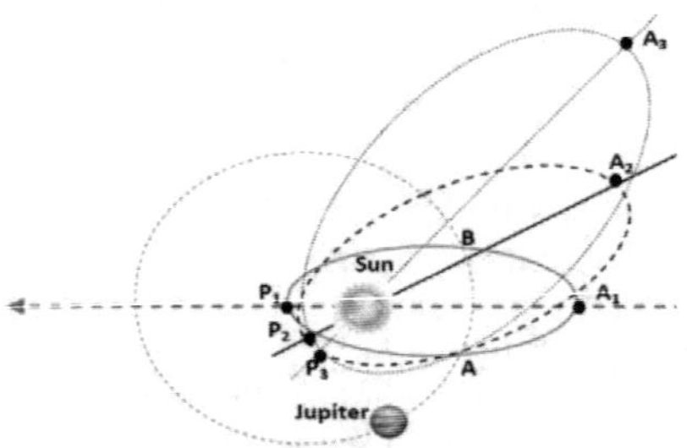

Figure 1: Overview of the orbital characteristics for some maneuvers.

The velocity and orbital elements of the spacecraft are changed by the close approach with Jupiter. The orbital elements (a = semi-major axis, e = eccentricity), energy (E) and angular momentum (C) of the spacecraft before the encounter with Jupiter are obtained from the equations: $a = \frac{r_a + r_p}{2}$, $e = 1 - \frac{r_p}{a}, E = -\frac{\mu}{2a}, C = \sqrt{\mu a(1 - e^2)}$, where r_p is the periapsis of the orbit of the spacecraft around the Sun, r_a is the apoapsis of that orbit and μ is the gravitational parameter of the Sun.

It is possible to obtain the variations of energy and angular momentum from equation [3]. The numerical values are shown in the Tables available in the results part of the paper.

$$\Delta v = \vec{v}_0 - \vec{v}_1 = 2|\vec{v}_\infty| \sin\delta \tag{1}$$

$$\Delta E = E_+ - E_- = -2v_2 v_\infty \sin\delta \, \sin\psi \tag{2}$$

$$\Delta C = \frac{\Delta E}{\omega} \tag{3}$$

where ω is the angular velocity between the primaries, δ is half of the angle of deflection due to the close approach and E_-, E_+ are the energy before and after the maneuver, respectively. Finally, having determined the variation of energy and angular momentum after the maneuver, it is possible to obtain the semi-major axis and the eccentricity of the orbit after the close approach, by using the equations $a = -\frac{\mu}{2E}$ and $e = \sqrt{1 - \frac{C^2}{\mu\, a}}$.

Next, the distance that the spacecraft has to pass from Jupiter (r_{ap}) to achieve an orbit that has a larger value for the apoapsis (so, achieving the goal of making the spacecraft to pass by different positions in space), but also going to an orbit that is resonant with the orbit of Jupiter, such that a new encounter will occur, is calculated. To solve this problem a list of resonant orbits is made and then it is arranged in order of increasing values of the energy. Then, a search is made to find the value of the r_{ap} for each passage that makes the orbits of the spacecraft to follow this sequence of orbits. Then the analytical solution can be obtained from the theory of hyperbolic orbits [21]:

$$r_{ap} = \frac{\mu}{v_\infty^2}\left[\frac{1}{\sin(\delta)} - 1\right] \quad (4)$$

where $\delta = \mathrm{ArcCos}[\frac{1}{\sqrt{2}}\sqrt{\frac{A^2+2B(B+C)-A\sqrt{A^2-4C(B+C)}}{A^2+B^2}}]$, $A = \frac{\sin(2\pi+\beta)}{2}$, $B = \cos(2\pi+\beta)$ *and* $C = -\frac{\Delta E}{2v_2 v_\infty}$.

3. Tisserand's Criterion

The Tisserand's criterion was introduced by astronomer Francois Felix Tisserand. It is an equation in dimensionless coordinates that represents a quantity that should be conserved in a circular restricted three-body system. Thus, two observed orbiting bodies are possibly the same if they nearly satisfy the Tisserand's Criterion [2]:

$$\frac{1}{a_i} + 2\sqrt{a_i(1-e_i{}^2)}\cos i_i \approx \frac{1}{a_0} + 2\sqrt{a_0(1-e_0{}^2)}\cos i_0 \quad (5)$$

Where a_i, e_i and i_i are the orbital elements before the maneuver and a_0, e_0 and i_0 are the orbital elements after the maneuver. In this study, Tisserand's Criterion will be used to validate the resonant orbits found. Note that, since only planar orbits are considered, the values for the inclination before and after the close approach is zero (direct orbits) or 180 degrees (retrograde orbits).

3. Numerical Study

The spacecraft starts its motion in a given orbit around the Sun, specified by its periapsis and apoapsis distances. A maximum number of revolutions for Jupiter between two successive close approaches are specified. It generates a large number of potential orbits and the maneuvers required to reach them. The meaning of "potential orbits and maneuvers" is that some of the maneuvers obtained may require a value smaller than the radius of Jupiter as a periapsis distance for one of the passages, or an orbit around the Sun with periapsis below the surface of the Sun. Those maneuvers have to be excluded from a practical list of maneuvers. A number higher than five would generate orbits with too large periods, which has little practical interest and that would be too much influenced by other perturbations, which are not considered here. Other perturbations, in particular the gravitational effects of Saturn, are not added to the dynamical model of the present research, because the idea here is to study possibilities of orbits and not making final decisions on which orbit to use. Those perturbations can be compensated by low cost maneuvers, after the nominal trajectories are found, or added in deeper studies of the orbits selected in this first study. Table 1 shows the orbits, including the information of the number of periods of Jupiter before the next close approach, the equivalent number of orbits of the spacecraft, the period (in years) of the orbit of the spacecraft, the respective semi-major axis (km), and the order of the resonance.

It is then possible to ordinate these orbits in crescent values of the energy and then search for values of r_{ap} for each passage that can make the spacecraft to follow this series. The following assumptions are considered: the passage occurs at the points A_n (Fig. 1); there are no other forces perturbing the spacecraft; the Sun-spacecraft energy and angular momentum are constant after and before the close approach.

Table 2 shows several different options for the first orbit. The possible orbits, after the first one, are shown in Table 3, that list the correct sequence of maneuvers that keeps the spacecraft increasing energy, semi-major axis and apoapsis distance. This Table shows the number of the maneuver, the orbital period (days), the distance of the closest approach obtained from the algorithm (in units of radius of Jupiter), semi-major axis (km), eccentricity, energy (km^2/s^2), periapsis distance (km), apoapsis distance (km), half of the deflection angle (degree), angle of approach (degree), order of the resonance and the time elapsed since the start of the maneuvers (years). The r_{ap}, the distance of the close approach between the spacecraft and Jupiter, is the key factor of the sequence, because it is assumed to be the only variable available to control the motion of the spacecraft. The radius of Jupiter is called R_J.

Now, it is necessary to study the second restriction and verify if the values of r_{ap} are possible to reach, which means that they have to be larger than the radius of Jupiter. Figures 2 to 5 show the results. Note that after maneuver 8, the distance of the closest approach is always below the surface of Jupiter. The times that the spacecraft remains in each orbit are visible in the plots, because the dots represent the instant of each maneuver.

Table 1: Resonant orbits for the spacecraft passing by Jupiter.

Revolutions of Jupiter	Revolutions of the spacecraft	*Period* of the spacecraft (years)	Semi-major axis of the spacecraft x 10^8 (km)	Order of resonance
1	1	11.859	7.785	1:1
	2	5.9295	4.904	2:1
2	1	23.718	12.358	1:2
	3	7.90599	5.941	3:2
	5	4.7436	4.226	5:2
3	1	35.577	16.193	1:3
	2	17.7885	10.201	2:3
	4	8.89424	6.426	4:3
	5	7.11539	5.538	5:3
	7	5.08242	4.425	7:3
	8	4.44712	4.048	8:3
4	1	47.436	19.615	1:4
	3	15.812	9.431	3:4
	5	9.48719	6.709	5:4
	7	6.77657	5.361	7:4
	9	5.27066	4.534	9:4
5	1	59.295	22.76	1:5
	2	29.6475	14.341	2:5
	3	19.765	10.94	3:5

The initial orbit for the spacecraft has periapsis of 155,000,000 km, which is near the orbit of the Earth around the Sun, and apoapsis of 1,011,830,000 km, that is higher than the orbit of Jupiter around the Sun. In that sequence it was not found the problem of reaching values for the periapsis that are inside the Sun. Some of the orbits (the first 4) have periapsis below the initial value, so those orbits intercept the orbit of the Earth around the Sun. It causes a potential risk of collision, but with very small probability. This possibility is neglected in the present study. In fact, these periapsis distances are interesting, because they generate trajectories that cross the path of the Earth and that can be used to recover the spacecraft, if desired. Note that the values for the distances of the closest approach decreases, to compensate the increase of the velocity of approach in the equation of the gain of energy (Eq. 2) and this fact limit the number of revolutions before a value below the surface of Jupiter is found for r_{ap}. This situation is a characteristic of the sequence of orbits shown here, where the values of important elements like the angle of approach and the velocity of approach are given in advance. Of course situations that generate an escape orbit in the first approach can be found for Jupiter, but this is not the goal of this research.

The energy, angular momentum, semi-major axis and apoapsis distances increase after each close approach. This characteristic is forced by the goal of finding series of orbits. The energy goes from

-113.98km^2/s^2 before the first passage until -65.19 km^2/s^2 after the last one, in crescent steps. This variation in energy causes the semi-major axis to go from 583.41x10^6 km to 1020.15x10^6 km, which corresponds to a variation of the apoapsis distances from 1011.83 x10^6 km (above the orbit of Jupiter) to 1679.94 x10^6 km. The eccentricity has a decreasing sequence and the time span for this sequence is years.

Table 2 – Options for initial conditions for the initial orbit (maneuver 0) .

Man.	period (day)	$r_{ap}(R_J)$	a (10^6 km)	e	Energy	r_p (10^6 km)	r_a (10^6 km)	δ (deg)	Ψ (deg)	resonance	Time (years)	Tisserand's Criterion
0	2266.8	-	505.58	0.693	-131,5	155.0	856.16	27.34	372.61	-	0	1.3101
0	2533.6	-	544.49	0,715	-122,1	155.0	933.99	36,38	372,65	-	0	1.2931
0	2810.0	-	583.41	0.734	-113.98	155.0	1011.83	40.42	373.52	-	0	1.2550

Table 3 - Sequence of orbits performing close approaches with Jupiter.

Man.	period (day)	$r_{ap}(R_J)$	a (10^6 km)	e	Energy	r_p (10^6 km)	r_a (10^6 km)	δ (deg)	Ψ (deg)	resonance	Time (years)	Tisserand's Criterion
0	2810.00	-	583.41	0.734	-113.98	155.0	1011.83	40.42	373.52	-	0	1.2550
1	1856.39	11.49	442.54	0.867	-150.27	58.75	826.33	33.31	353.59	7:3	35.60	1.2551
2	2165.79	17.41	490.44	0.814	-135.59	91.3	889.57	43.73	356.00	9:4	83.00	1.2545
3	2475.19	9.47	536.10	0.771	-124.04	122.93	949.27	48.99	358.74	2:1	142.31	1.2544
4	2598.94	6.892	553.82	0.756	-120.07	135.08	972.57	52.70	357.55	7:4	177.89	1.2548
5	2887.71	5.448	594.12	0.727	-111.93	162.07	1026.18	57.55	357.60	5:3	201.60	1.2549
6	3248.68	3.920	642.66	0.700	-103.48	192.97	1092.34	61.14	358.81	3:2	237.18	1.2544
7	3465.19	3.004	670.9	0.687	-99.12	209.99	1131.81	65.93	356.40	4:3	284.16	1.2547
8	4331.57	2.018	778.52	0.656	-85.42	267.77	1289.27	73.25	356.27	5:4	296.48	1.2547
9	5775.33	0.938	943.10	0.644	-70.51	335.64	1550.56	78.28	358.70	1:1	343.91	1.2547
10	6497.36	0.451	1020.15	0.647	-65.19	360.36	1679.94	80.65	358.93	3:4	379.49	1.2544

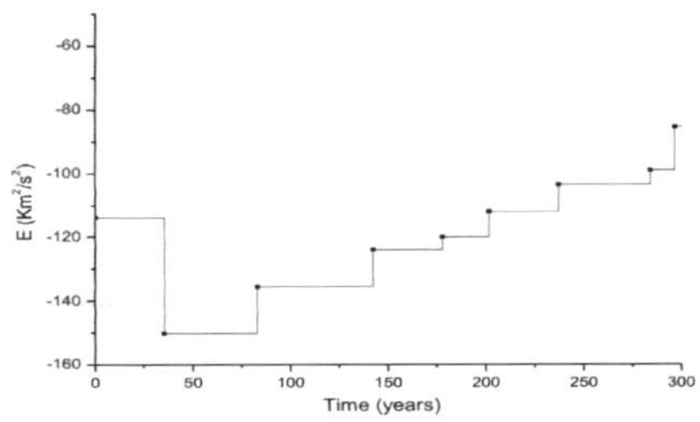

Figure 2: Energy of the spacecraft as a function of time.

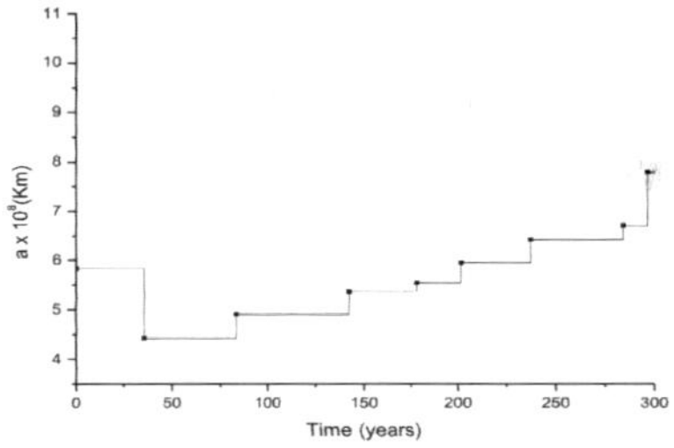

Figure 3: Semi-major axis of the spacecraft as a function of time.

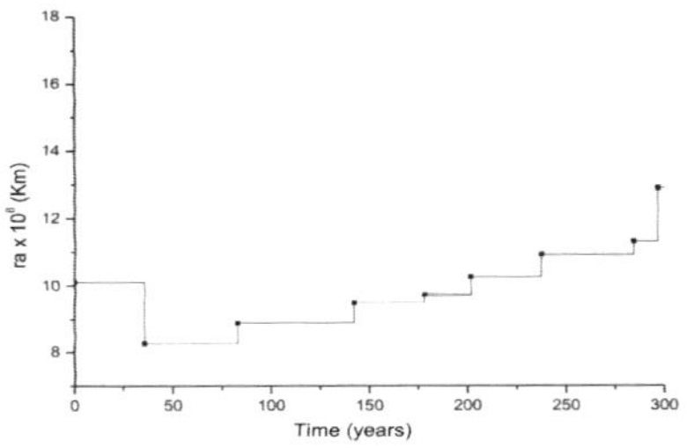

Figure 4: Apoapsis distance of the spacecraft as a function of time.

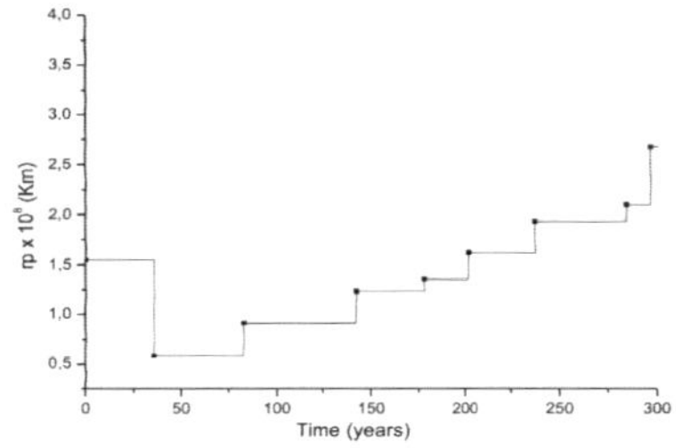

Figure 5: Periapis distance of the spacecraft as a function of time.

4. Conclusion

This study was made to show the evolution of the trajectories, as well as the amplitudes of the variations of the keplerian elements, velocity, energy and angular momentum in a trajectory of a spacecraft that perform a series of close approaches with the planet Jupiter. A set of analytical equations is used to allow the calculation of the distance of the closest approach that generates a specified orbit. Then, a series of resonant orbits with Jupiter that is increasing apoapsis distance to cover a large area of the space near the orbit of Jupiter is found. Using these equations it is possible to establish a sequence of close approaches that meets the goals. The results showed that it is possible to find useful sequences of close approaches to study the space near the orbit of Jupiter by using these natural changes of orbits to pass by different positions in the space without the expenses of applying a control to the spacecraft.

Acknowledgment

The authors wish to express their appreciation for the support provided by grants # 473387/2012-3 and 304700/2009-6, from the National Council for Scientific and Technological Development (CNPq); grants # 2011/09310-7 and 2011/08171-3, from São Paulo Research Foundation (FAPESP) and the financial support from the National Council for the Improvement of Higher Education (CAPES).

References

[1] Allen J AV 2003 *Amer. Jour. Phys.* **71** 448-4515
[2] Battin R H 1987 *An introduction to the mathematics and models of astrodynamics AIAA*, (New York: New York)
[3] Broucke R A 1988 The Celestial mechanics of gravity assist, *In:AIAA/AAS Astrodynamics Conf., Minneapolis*, 88
[4] Prado A F B A and Broucke R A 1995 *Jour. Guid. Cont. and Dyn.* **3**(5) 593-598
[5] Prado A F B A and Broucke R A 1996 *Jour. Guid. Cont. and Dyn.* **19**(4) 929-933
[6] Prado A F B A 1996 *Acta Astron.* **39** 483
[7] Flandro G 1996. *Astr. Acta*, **12** 4
[8] D'Amario L A, Byrnes D V and Stanford R H 1982 *Jour. Guid. Cont. and Dyn.* **5** 465-471
[9] Dowling R L, Kosmann W J, Minovitch M A. and Ridenoure R W 1991 Gravity Propulsion Research at UCLA and JPL *in: 41st Con. of the IAF* (Dresden, GDR, Oct. 1991) pp 6-12
[10] Farquhar R W and Dunham D W 1981 *Jour. Guid. Cont. and Dyn.* **4**(2) 192-196
[11] Striepe S A and Braun R D 1991 *Journal of the Astronautical Sciences* **39**(3) 299-312
[12] Swenson B L 1992 Neptune Atmospheric Probe Mission *AIAA in: AIAA7AAS Astr. Conf.* (Hilton Head: SC, 10-12 August 1991) 92-4371
[13] Weinstein S S 1992 Pluto Flyby Mission Design Concepts for Very Small and Moderate Spacecraft in: AIAA/AAS Astr. Conf. (Hilton Head, SC, Aug. 10-12) 92-4372
[14] Petropoulos A E and Longuski J M 2000 Automated design of low-thrust gravity-assist trajectories *in AIAA/AAS Astrodynamics Specialist Conference (*Denver,CO, August 14-17) 2000-4033
[15] Petropoulos A E, Longuski J M and Vinh N X 1999 Shape-based analytic representations of low-thrust trajectories for gravity-assist applications *in: AAS/AIAA Astr, Conf.* (Girdwood, Alaska, Aug. 16-19) 99-337
[16] Strange N J and Longuski J M 2002 *Jour. of Spac. and Rock.* **39** 9-16
[17] Prado A F B A 1997 *Jour. Guid. Cont. and Dyn.* **19** 1142-47
[18] Prado A F B A 1997 *Jour. Guid. Cont. and Dyn.* **20** 797-802
[19] Gomes V M and Prado A F B A 2010 A Study of the impact of the initial energy in a close approach of a cloud of particles *WSEAS Trans. on Math.* **9** 811-820
[20] Santos D P S, Prado A F B A and Roco E M 2013 Math Prob in Eng, **2013**, 1- 13
[21] Formiga J K S. and Prado A F B A 2011 Orbital Characteristics Due to the Three Dimensional Swing-By in the Sun-Jupiter System *in: 10 International Conference on Computational Intelligence Man-Machine Systems and Cybernetics, Jakarta*: *The World Scientific and Engineering Academy and Society Press*, **1**. 61-69
[22] M Hénon 1968 *Bulletin Astronomique* **3** 377
[23] K C Howell 1987 *Cel. Mech.* **40** (3-4) 393–407

Semi-analytical study of the rotational motion stability of artificial satellites using quaternions

Josué C. dos Santos[1], Maria Cecília Zanardi[2], Nicholas Matos[3]
São Paulo State University, Group on Orbital Dynamics and Planetology, Guaratinguetá, Brazil
E-mail: [1]josuesantosunesp@gmail.com, [2]cecilia@feg.unesp.br, [3]nicholas.f.matos@hotmail.com

Abstract.
This study at aims performing the stability analysis of the rotational motion to artificial satellites using quaternions to describe the satellite attitude (orientation on the space). In the system of rotational motion equations, which is composed by four kinematic equations of the quaternions and by the three Euler equations in terms of the rotational spin components. The influence of the gravity gradient and the direct solar radiation pressure torques have been considered. Equilibrium points were obtained through numerical simulations using the softwares Matlab and Octave, which are then analyzed by the Routh-Hurwitz Stability Criterion.

1. Introduction

In this paper the stability analysis of the artificial satellite rotational motion is performed by applying the Routh-Hurwitz Criterion (RHC). This criterion allows the investigating of the static stability for systems of equations using the coefficients of the characteristic equation associated with the linearized system (in this paper, the system of equations for the rotational motion), without the need of determining the characteristic equation roots. Quaternions are used to describe the satellite spatial orientation (attitude) and the gravity gradient torque and the solar radiation torque are included in dynamic equations of rotational motion. These torques are discussed in section 3 and their components are determined using rotational matrices composed by the quaternion attitude. For the RHC application, the motion equations are linearized around each equilibrium point determined by an implemented program using the softwares Matlab and Octave. Applications are performed for satellites with similar characteristics to the American satellite PEGASUS and to the Brazilian satellites SCD-1 and SCD-2. These Brazilian satellites are spin-stabilized and the rotation axis is aligned in the direction of the greatest satellite principal moment of inertia axis. The RHC checks the static stability of motion, no matter how it's motion, but whether it oscillates in the desired condition. The eigenvalues from the Jacobian matrix associated to the linearized system are also obtained in order to verify if the real parts of eigenvalues were nulls or negatives and if the imaginary parts were not zero. In order to investigate the range around the equilibrium points, there are introduced small variations in the principal moments of inertia of the satellite. In this paper is assumed that the satellite is illuminated all the time.

2. Attitude Representation and Motion Equations

2.1. Quaternion Attitude

Quaternions were introduced by Hamilton in 1843 and the main idea was introduced as the extension of the complex numbers. They form a non-commutative algebra in the space $\Re^4$. Nowadays, quaternions are used in several research areas like Theory of Numbers, Theorical Physics, High Energy Physics, Differential Geometry, Applied Mathematics and Aerospace Engineering. In the context of the space research, they can be used to represent the spatial rotations in a compact and convenient form. For more details, see [2, 5, 13]. If Φ is the rotation angle and $\begin{pmatrix} n_1 & n_2 & n_3 \end{pmatrix}^t$ is the unit vector aligned to the rotation axis, with t indicating the transposed matrix, the rotation quaternion is a vector (4x1) defined by [9, 13]

$$Q = (\vec{Q} \;\; q_4)^t = (q_1 \;\; q_2 \;\; q_3 \;\; q_4)^t \tag{1}$$

with $q_1 = n_1 \sin(\Phi/2)$, $q_2 = n_2 \sin(\Phi/2)$, $q_3 = n_3 \sin(\Phi/2)$ and $q_4 = \cos(\Phi/2)$, where $\vec{Q}=\begin{pmatrix} q_1 & q_2 & q_3 \end{pmatrix}$ is the quaternion vectorial component and q_4 the scalar quaternion component. The main properties of quaternions are its unitary module and the attitude matrix corresponding to a rotation Φ about the axis $\hat{n} = \begin{pmatrix} n_1 & n_2 & n_3 \end{pmatrix}^t$. This matrix can be seen in [9, 15]

2.2. Kinematic and Euler Equations

The rotational motion equations are given by the dynamic and kinematic equations. The dynamic equations for the artificial satellite rotational motion are described on satellite principal-axis-of-inertia system (Oxyz) which is called in this paper as Rotating System and is given by [9, 15]

$$I_x \, \dot{p} = N_x + (I_y - I_z)qr \tag{2}$$

$$I_y \, \dot{q} = N_y + (I_z - I_x)pr \tag{3}$$

$$I_z \, \dot{r} = N_z + (I_x - I_y)pq \tag{4}$$

where I_x, I_y e I_z are the principal moments of inertia of the satellite; p, q e r are the spin velocity components on Rotating System; N_x, N_y e N_z are the resulting torque components due to the external torques on Rotating System. The time dependence for the quaternion components are given by the kinematic equations [11]

$$\dot{q_1} = 0.5[pq_4 - qq_3 + rq_2] \tag{5}$$

$$\dot{q_2} = 0.5[qq_4 - rq_1 + pq_3] \tag{6}$$

$$\dot{q_3} = 0.5[rq_4 - pq_2 + qq_1] \tag{7}$$

$$\dot{q_4} = -0.5[pq_1 - qq_2 + rq_3] \tag{8}$$

By theses equations there are not any singularities associated to the null denominator or indefinitions in the variables.

3. External torques

3.1. Gravity Gradient Torque

The gravitational force on one part of the spacecraft is different from the another, because the magnitude of the Earth gravitational force varies with the inverse of the square of the distance from the geocenter and this difference result in a torque called gravity gradient torque (GGT). The GGT model adopted in this paper is a classic one used in previous papers. Some applications were developed like analytical propagation of the rotational motion with Andoyer variables [12,14] and numerical propagation with the quaternions [15]. The GGT is also included in stability analysis [7, 12].The GGT model [9, 13] for a circular cylinder shape satellite is given by

$$\vec{N_G} = 3\frac{\mu}{r'^3}[a_{21}a_{31}(I_z - I_y)\vec{i} + a_{11}a_{31}(I_x - I_z)\vec{j} + a_{11}a_{21}(I_y - I_x)\vec{k}] \tag{9}$$

where the elements a_{11} , a_{21} e a_{31} are given in terms of the attitude quaternions and orbital angular elements [10]. Therefore, if the satellite has a homogeneous mass distribution (equipments on board) or a spherical symmetry around of the center of mass (which is equivalent to $I_x = I_y = I_z$), the influence of this torque is null; otherwise, its influence becomes significant. The GGT can be used for the stabilization of satellites, for which it is necessary a good precision in its spatial orientation. This is possible due to the existence of two stable equilibrium points, which are lagged in an angle of 180 ° [3]. This torque decreases with the cube of altitude, making the gravity gradient stabilization of satellites in high altitude orbits unviable[1], due to the small magnitude of this torque.

3.2. Solar Radiation Torque

The Sun and the Earth are the most important radiation sources in space. The provided radiation from the Earth has two significant portions: the solar radiation reflected diffusely on the Earth surface (terrestrial albedo), and the terrestrial radiation, that is the spontaneous emission in the infrared range, proportional to the fourth power of the absolute terrestrial surface temperature. The albedo and the terrestrial radiation are respectively 90% and 93% smaller (at 700 km of altitude) than the direct solar radiation (at 700 km of altitude). Due to their magnitudes, we can ignore their contributions and then we can consider only the direct solar radiation as a good approximation for the total solar radiation received by the satellite surface. As the satellite-Sun distance is much greater than the satellite-Earth distance, this torque is practically independent of the satellite altitude for the satellites in little geocentric altitude [13].The solar direct radiation pressure force is created by the continuous photon collisions with the satellite surface. The satellite surface can be able to absorb or to reflect this flow. The direct solar radiation pressure force arise from the total change of the momentum of all the incident photons on the satellite surface and it can produce a torque known as direct solar radiation pressure torque (DSRPT). It can cause considerable disturbances in spacecraft orbits when the satellite has certain characteristics such as great ratio, high altitude and great solar panels. The model for the DSRPT as developed by [14] has z component null ($\bar{N}_z = 0$) and the x and y components are

$$\bar{N}_x = -\frac{\bar{K}}{R^4}(\beta_1\gamma_1 - \beta_2\gamma_2)\frac{h}{2}\pi\sigma^2\{a_s{}^2\, R_y\, R_z + a_s\, r'(R_y\, a_{31} + R_z\, a_{21}) + r'^2\, a_{21}\, a_{31}\} \tag{10}$$

$$\bar{N}_y = \frac{\bar{K}}{R^4}(\beta_1\gamma_1 - \beta_2\gamma_2)\frac{h}{2}\pi\sigma^2\{a_s{}^2\, R_x\, R_z + a_s\, r'(R_x\, a_{31} + R_z\, a_{11}) + r'^2\, a_{11}\, a_{31}\} \tag{11}$$

where $\bar{K}$ is a solar parameter and assume the value[2] 1.01x10^{17}kgm/s, a_s is Sun-Earth distance and here assume the value 1.49597870x10^{11}m, r' is the satellite geocentric distance, R_s is the Sun-satellite distance, β_i e γ_i, i=1,2, are the specular and total reflection coefficients, respectively, for each satellite surface (which assume constant values), a_{11}, a_{21} e a_{31} are direction cosines which relate the Orbital System and the Satellite System (which are composed by the quaternions [10]) and R_x, R_y and R_z give the Sun direction in the Satellite System.

[1] For example: the geostationary orbits

[2] According G M Georgevic:"The Solar Radiation Pressure Force and Torques Model", The Jour. Astron. Scienc., XX, 5, p. 257 (1973).

4. Algorithm for Stability Analysis

The chosen criterion for a preliminary stability analysis of the rotational motion is the Routh-Hurwitz Criterion (RHC) [4, 8], which allows for investigating the systems of equations static stability using the coefficients of the characteristic equation associated with the linearized system, without the need to determine the roots of this characteristic equation. For applying the RHC is required the linearization of the system of motion equations, which corresponds here, the three dynamic equations in terms of spin components, where are included GGT and DSRPT components, and the four kinematic equations related to the four quaternion attitude components [15]. The procedure for RHC is described below: 1) Write the characteristic equation for the linearized system of equations as follow [10]

$$a_n\,\lambda^n + a_{n-1}\,\lambda^{n-1} + ... + a_2\,\lambda^2 + a_1\,\lambda + a_0 = 0 \tag{12}$$

In this paper, the coefficients a_l , l = n, n - 1, ... , 0 are dependent on the satellite moments of inertia, the semimajor axis of the orbit, the longitude of the ascending node, the argument of pericenter, the orbital inclination, and initial conditions for the attitude quaternions, the true anomaly and the spin velocity. 2) If one coefficient is zero or negative in the presence of another positive coefficient, then there is at least one root with positive real part, and then the system is not stable [4]. 3) If all coefficients are positive, must be constructed a table (Routh table), whose elements are associated with the coefficients of the characteristic equation in rows and columns: with a_{n+2-2j} e a_{n+1-2j} being the characteristic equation coefficients[3] [10]. In

j / k = i - 1		1	2	3	...	j
0	λ^n :	a_n	a_{n-2}	a_{n-4}	....	a_{n+2-2j}
1	λ^{n-1} :	a_{n-1}	a_{n-3}	a_{n-5}	....	a_{n+1-2j}
2	λ^{n-2} :	b_{n-2}	b_{n-4}	b_{n-6}	....	b_{n-2j}
3	λ^{n-3} :	c_{n-3}	c_{n-5}	c_{n-7}	....	c_{n-1-2j}
⋮	⋮ ⋮		⋮	⋮		⋮
k	λ^{n-k} :	z_{n-k}	z_{n-k-2}	z_{n-k-4}	....	$z_{n-k+2-2j}$
⋮	⋮ ⋮		⋮	⋮		
n - 4	λ^4 :	α_4	α_2	α_0		
n - 3	λ^3 :	β_3	β_1			
n - 2	λ^2 :	γ_2	γ_0			
n - 1	λ^1 :	δ_1				
n	λ^0 :	ε_0				

Figure 1. Scheme for the Routh table general form. [10]

this paper are considered seven variables (the four quaternions components and the three spin velocity components) and the Routh table for the case n=7 variables is represented in the scheme in Figure 2 [10].

5. Results obtained by numerical simulations

The RHC was applied to analyze the rotational motion stability of three satellites: 1) A medium sized satellite (MS) with similar data[4] to the north-American satellite PEGASUS. 2) Two small sized satellite (SS-1 e SS-2) with similar data to the Brazilian Data Collection satellites, SCD1 and SCD2. Simulations were performed for each satellite, using the softwares Matlab and Octave, considering different initial conditions for spin velocity and quaternions and for the satellites in

[3] The Routh table elements can be seen with details in [10]

[4] These satellites data can be seen in [10]

λ^7 : a_7 a_5 a_3 a_1
λ^6 : a_6 a_4 a_2 a_0
λ^5 : b_5 b_3 b_1
λ^4 : c_4 c_2 c_0
λ^3 : d_3 d_1
λ^2 : e_2 e_0
λ^1 : f_1
λ^0 : g_0

Figure 2. Scheme for the Routh table in the case n=7 variables. [10]

Satellite	10^{-5}	10^{-6}	10^{-7}	10^{-9}	Total
MS	9	2	2	1	14

Table 1. Order of magnitude of the eigenvalues with real part less then 10^{-5}.

Real Part	Imaginary Part
$1.251\text{x}10^{-4}$	$1.610\text{x}10^{-3}$
$1.251\text{x}10^{-4}$	$-1.610\text{x}10^{-3}$
$1.060\text{x}10^{-3}$	0
$-7.029\text{x}10^{-4}$	$7.373\text{x}10^{-4}$
$-7.029\text{x}10^{-4}$	$-7.373\text{x}10^{-4}$
$4.800\text{x}10^{-5}$	$7.448\text{x}10^{-4}$
$4.800\text{x}10^{-5}$	$-7.448\text{x}10^{-4}$

Table 2. Eigenvalues associated to the MS equilibrium point x_0.

different positions around their orbits (values for the true anomaly and the initial rotation angle Φ were 0°, 90°, 180° and 270°). The results did not show any stable equilibrium points by the RHC. However, analyzing the eigenvalues associated to the obtained equilibrium points, it is possible to observe many eigenvalues with a very small order of magnitude in their real part (values less than 10^{-5}). These ones produced solutions with small changes during long times[5]. There were obtained 14 points for MS, 56 points for SS1 and 49 points for SS2 with theses characteristics. The results for the satellite MS are presented in Table 1.

The values for the first column of the Routh table for each equilibrium point were obtained by numerical simulations. To illustrate this application for the MS equilibrium point x_0=(p,q,r,q_1,q_2,q_3,q_4)=($6.534\text{x}10^{-4}$, $6.121\text{x}10^{-4}$, $6.140\text{x}10^{-4}$, $6.019\text{x}\ 10^{-1}$, $9.001\text{x}10^{-2}$, $9.001\text{x}10^{-1}$, $6.079\text{x}10^{-12}$), we obtained the following first column elements: a_7=1.000, $a_6=4.499\text{x}10^{-18}$, $b_5=-2.281\text{x}10^{7}$, $c_4=-2.290\text{x}\ 10^{-10}$, $d_3=1.133\text{x}10^{-10}$, $e_2=-2.690\text{x}10^{-15}$, $f_1=3.190\text{x}10^{-19}$ and $g_0=-1.597\text{x}10^{-21}$. The eigenvalues for this equilibrium point are presented in Table 2, where we observed the small order for the eigenvalues real part(see footnote 5).

In order to investigate the range around the equilibrium points, there are introduced small variations in the principal moments of inertia of the satellite [6]. For each initial moment of inertia (I_{x0},I_{y0},I_{z0}) were introduced variation between $0.9I_{x0}$ and $1.1I_{x0}$, $0.9I_{y0}$ and $1.1I_{y0}$ and

[5] Eigenvalues with positives real part equal to 10^{-4}, 10^{-5}, 10^{-6}, 10^{-7} and 10^{-9} produce an increase of approximately 2.7 times in periods of 2.78 hours, 27.80 hours, 11.60 days, 3.86 months and 3.17 years respectively

$0.9 I_{z0}$ and $1.1 I_{z0}$. There were investigate 305 equilibrium points in order to get RHC stable points. Various RHC stable points were obtained: 647 for MS, 297 for SS-1 and 647 for SS-2. For example, for the MS satellite with I_{x0}= 3.1599x10^5 Kg.m^2 and I_{y0}=3.6461x10^5 Kg.m^2, we have three RHC equilibrium points with I_{z0}=9.2754x10^4 Kg.m^2, 1.0175x10^5 Kg.m^2 and 1.1071x10^5 Kg.m^2.

6. Conclusions

In this paper is presented an approach to analyze the stability of rotational motion by the Routh-Hurwitz Criterion. Many simulations were developed and RHC stable regions were obtained around the equilibrium points. The goal of this paper is the use of quaternions in determining the DSRPT and GGT components on Rotating System for the stability analysis. This approach is valid for cylindrical satellites in the illuminated phase of the trajectory and can be useful for analyzing the stability of future missions of the Brazilian space program.

However due the equations of rotational motion are non linearized and others stability analysis methods could be applied in order to get better results. We can highlight among them the Phase Space Method, the Liapunov Direct Method and the Kovalev-Savchenko Theorem [7,12]. Others approachs could include the Earth's shadow in the DSRPT and also include the others external torques.

7. Acknowledgments

The authors are wishing to acknowledge support from UNESP, CNPq and FAPESP (process n° 2012/21023-6). Special thanks to God for all the opportunities given during the development of this paper.

References

[1] Creanshaw J W and Fitzpatrick P M 1968 *AIAAJ* **6** 2140
[2] Kane T R, Likins W, Levinson D A 1983 *Spacecraft dynamics*(McGraw Hill Book Company, New York)
[3] Kaplan M H 1976 *Modern spacecraft dynamics and control*(John Wiley & Sons, New York)
[4] Kuo B C 1985 *Sistemas de Controle Automático*(Prentice Hall, Rio de Janeiro)
[5] Lindegren L, Lammers U, Hobbs D, Mullane W O, Bastian U, Henández J 2011 *[arXiv:1112.4139]*
[6] Matos N F O 2012 *CNPq Undergraduate Research Final Report*(UNESP, Guaratinguetá)
[7] Moraes R V, Cabette R E S, Zanardi M C, Stuchi T J, Formiga J K 2009 *Celestial Mechanics & Dynamical Astronomy* **104** 4 pg.337-353
[8] Ogata K 2009 *Modern Control Engineering*(Prentice Hall, New York)
[9] Pisacane V L and Moore R C 1994 *Fundamentals of Space Systems*(Oxford University Press, New York)
[10] Santos J C 2011 (UNESP, Guaratinguetá) [Eletronic adress: *http://acervodigital.unesp.br/handle/123456789/64610*]
[11] Shuster M D 1993 *Journ. of Astron. Scienc* **41** 4 pg. 439-517
[12] Silva W R, Zanardi M C, Cabette R E S, Formiga J K S 2012 *Matem. Prob. in Engin.* **2012** 137672
[13] Wertz J R 1978 *Spacecraft attitude determination and control* (Dordrecht, The Netherlands)
[14] Zanardi M C and Moraes R V 1999 *Celestial Mechanics & Dynamical Astronomy* **75** pg.227-250
[15] Zanardi M C 2005 *Dynamics of artificial satellite*(UNESP, Guaratinguetá)

Orbital Disturbance Analysis due to the Lunar Gravitational Potential and Deviation Minimization through the Trajectory Control in Closed Loop

L D Gonçalves[1], E M Rocco[1] and R V de Moraes[2]

[1] Instituto Nacional de Pesquisas Espaciais, INPE, São José dos Campos, Brazil
[2] Universidade Federal de São Paulo, UNIFESP, São José dos Campos, Brazil

E-mail: lianadgon@gmail.com, evandro_mr@yahoo.com.br, rodolpho.vilhena@gmail.com

Abstract. A study evaluating the influence due to the lunar gravitational potential, modeled by spherical harmonics, on the gravity acceleration is accomplished according to the model presented in Konopliv (2001). This model provides the components x, y and z for the gravity acceleration at each moment of time along the artificial satellite orbit and it enables to consider the spherical harmonic degree and order up to100. Through a comparison between the gravity acceleration from a central field and the gravity acceleration provided by Konopliv's model, it is obtained the disturbing velocity increment applied to the vehicle. Then, through the inverse problem, the Keplerian elements of perturbed orbit of the satellite are calculated allowing the orbital motion analysis. Transfer maneuvers and orbital correction of lunar satellites are simulated considering the disturbance due to non-uniform gravitational potential of the Moon, utilizing continuous thrust and trajectory control in closed loop. The simulations are performed using the Spacecraft Trajectory Simulator-STRS, Rocco (2008), which evaluate the behavior of the orbital elements, fuel consumption and thrust applied to the satellite over the time.

1. Introduction

If the existence of disturbing forces were ignored, the orbital motion would be a conic, set in a fixed plane, with constant size and eccentricity. However, the existence of such forces tends to cause variations in the elements that characterize the orbit of an artificial satellite. In some cases this variations should be corrected to enable the mission accomplishment.

In order to study the perturbations due to non-sphericity of the lunar gravitational field it is used the LP100K model, so that an analysis of the influence of the degree and order of the harmonics in the artificial lunar satellite orbit. is done. These disturbance effects are inserted into the Spacecraft Trajectory Simulator (STRS) in order to control the trajectory and minimize the deviations. The correction of the errors in the orbit is made by the STRS using a continuous propulsion system controlled in closed loop.

2. Lunar Gravitational Potential

The moon's gravitational potential is expressed by the coefficients of normalized spherical harmonics, given by Equation (1) (Konopliv, 2001; Kuga, 2011):

$$U(r,\lambda,\phi) = \frac{\mu}{r} + \frac{\mu}{r}\sum_{n=2}^{\infty}\sum_{m=0}^{n}\left(\frac{a_e}{r}\right)^n (\bar{C}_{nm}\cos m\lambda + \bar{S}_{nm}\,\mathrm{sen}\, m\lambda)\,\bar{P}_{nm}\,(\mathrm{sen}\,\phi) \tag{1}$$

where n is the degree, m is the order, μ is the gravitational constant and r is the lunar equatorial radius. $\bar{P}_{nm}$are the fully normalized associated Legendre polynomials. a_e is the reference radius Moon, $\emptyset$ is the latitude, and λ is the longitude.

3. The Model LP100K

The lunar gravitational field was determined using data from some previous lunar missions. One of the most important missions was the Lunar Prospector (1998-1999). The LP was the third mission in NASA's exploration program, called Discovery, and provided the first measurement of the moon gravitational field. The information about the gravity field comes from the long-term effect observed in the satellite orbit (Konopliv, 2001).

The model presented in Konopliv's called GRAVITYSPHERICALHARMONIC is a representation of the spherical harmonics due to planetary gravity, based on the gravitational potential of the planet, given by the Equation (1).

The output calculated by the model includes the values of gravity in meters per squared second, on the axes x, y and z. From the values of the gravity acceleration it is possible to obtain the state variables and hence the orbital elements that characterize the satellite orbit.

Using the GRAVITYSPHERICALHARMONIC model it was created the Gravity_Moon subroutine, used for the simulations of the artificial satellites orbital motion around the Moon's surface.

4. Study of oblateness and equatorial ellipticity effects on the lunar orbit of an artificial satellite

The Figure 1 shows the value obtained for the resulting of gravity acceleration on a satellite for each value of degree and order from 1 up to 100, at an altitude around 250 km.

We observe, by the Figure 1 that the value of the gravity acceleration on the artificial satellite tends to stabilize at a value close to 1.2250 m/s², when considering values of degree and order bigger than 15. However, we can also reach this approximate value using the degree and order 2.

It is important to adopt the highest possible value for degree and order, since by the use of many terms of the spherical harmonics we can represent the imperfections of the bodies format in a more accurate way. However, for a first analysis, the value 2 for degree and order could be adopted.

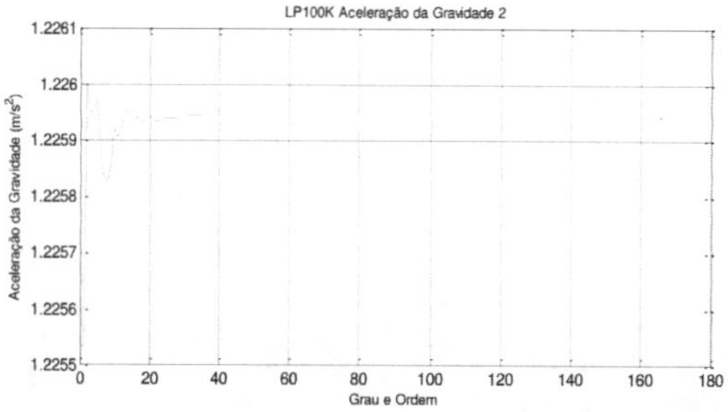

Figura 1. Gravity acceleration due spherical harmonics

5. Results

In this section we present the results of two different simulations, both performed during 86400 s, considering the terms due to non-homogeneity of the lunar gravitational field, and the value 2 to degree and order. In both cases, the continuous propulsion system are trigged when the simulation time reaches 2000 s and is turned off when the semi-major axis reaches the value of 4000 km.

In the first simulation the satellite leaves a low lunar orbit to reaches a high orbit, using continuous tangential thrust with magnitude of 2N, as seen in Figure 2. In the second simulation it is considered the application of higher thrust (20 N), applied over an arc of 5 degrees around the periapse, as seen in Figure 3. In the simulations 1 and 2 were considered the effects of disturbance and the action of the thrusters simultaneously, whose initial conditions considered were: semi-major axis 1800000 m; eccentricity 0.001; inclination 45°; right ascension of the ascending node 20°; argument the periapse 100°; mean anomaly 1°.

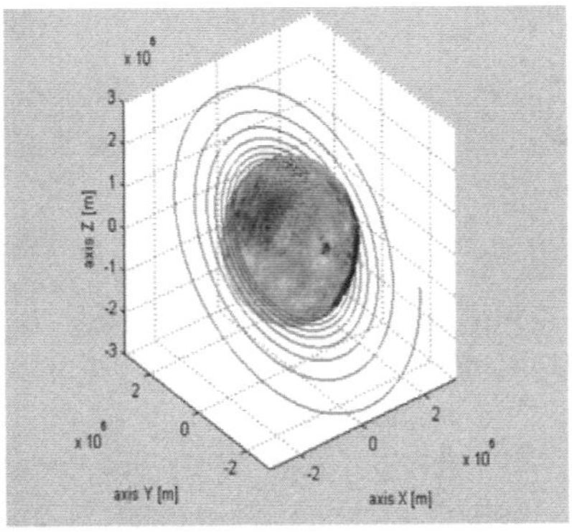

Figure 2. Trajectory of the satellite in simulation 1

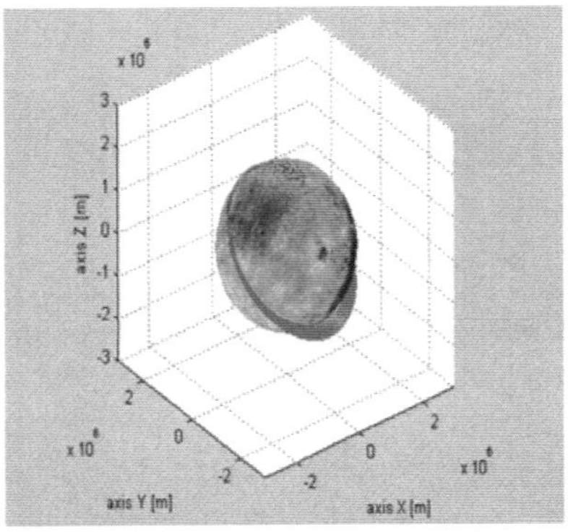

Figure 3. Trajectory of the satellite in simulation 2

The Figure 4 presents the case where only the correction of the trajectory is considered to illustrate the ability of the control system to deal with the effect of orbital perturbation. In this case, the aim of the control system is minimize the effects of the perturbations acting on the satellite. In this simulation only the effects of lunar oblateness and equatorial ellipticity were considered until degree 2. From Figure 4 we can notice that the force applied by the propulsion system acts toward to correct the effects caused by the disturbing force. The figure shows the results for x axis, but a similar behavior is obtained for y and z axes. Therefore, it was verified that the control system is able to deal with the disturbance effects on the lunar satellite when considering the effects caused by the non-homogeneity of the lunar gravitational field.

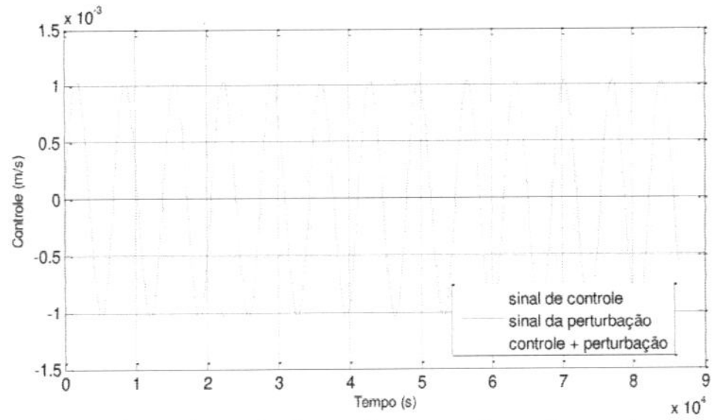

Figure 4 - Control signal and disturbance signal on the satellite (x axis)

The following results obtained in the simulations 1 and 2 will be exposed to the study of the behavior of the orbital elements, propellant mass, thrust applied on the satellite, altitude reached and the disturbance acting on the satellite along the trajectory.
The Figures 5 and 6 show the behavior of the semi major axis in the two simulations. The Figure 5 shows the variation of the semi-major axis during the orbital maneuver. In the Figure 6 we can see that each propulsive arc produces a step, more exactly, each application of the arc causes a sudden increase of the semi-major. It is also observed that the application of each arc causes a deviation in actual trajectory of the satellite, characterized by a difference observed at the beginning of each step in Figure 6, when the propellant system is turned on. However, the control system operates to maintain the actual path close to the reference.

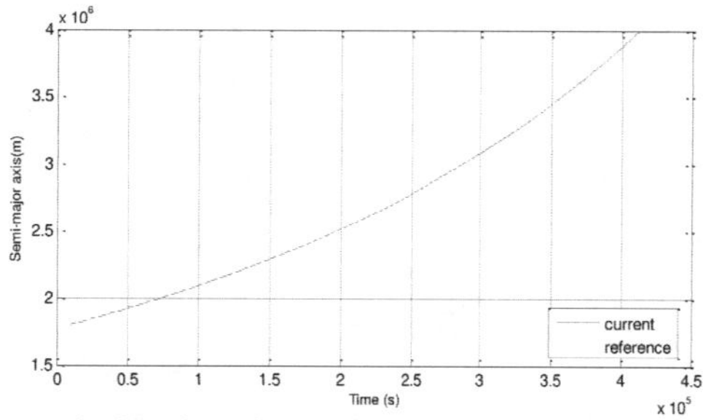

Figure 5. Semi-major axis of the satellite in the simulation

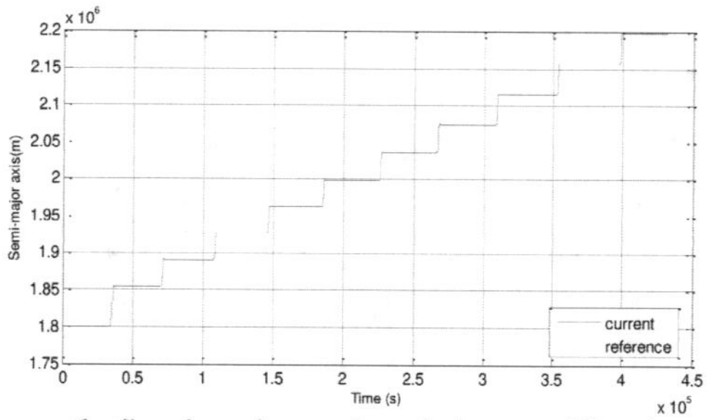

Figure 6. Semi-major axis of the satellite in simulation 2

In the Figures 7 and 8, we can verify that throughout the maneuver, the eccentricity presents small variations due to the applied thrust and the disturbance of the lunar gravitational potential. We can note that the eccentricity always tends to increase. However, in the Figure 7 we see that this value oscillates. The Figure 8 shows that application of the propulsive arcs always causes the increase of the eccentricity, but the eccentricity remains constant between the applications of the propulsive arcs.

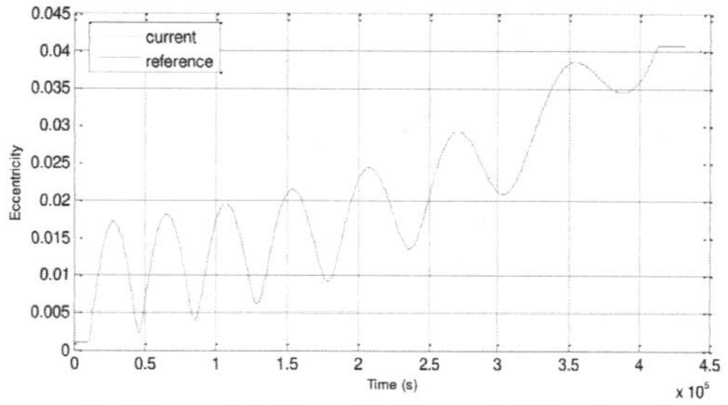

Figure 7. Eccentricity of the satellite in the simulation 1

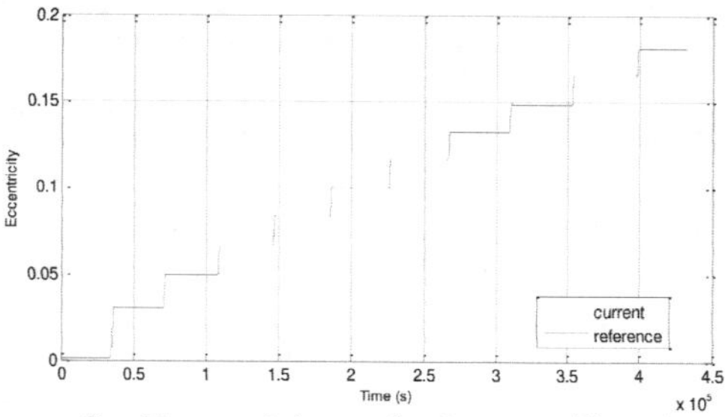

Figure 8. Eccentricity of the satellite in simulation 2

The altitude of the satellite in the simulations is shown in the Figures 9 and 10. The oscillations in the graphs are justified because the satellite's altitude varies during each orbit. When the satellite is in an apoapse, it is at a slightly higher orbit, and when the satellite is in periapse, is at a slightly lower orbit. We observe a considerable fluctuation in the result found in the second simulation since a thrust is applied on the satellite every time that the passage through periapse occurs.

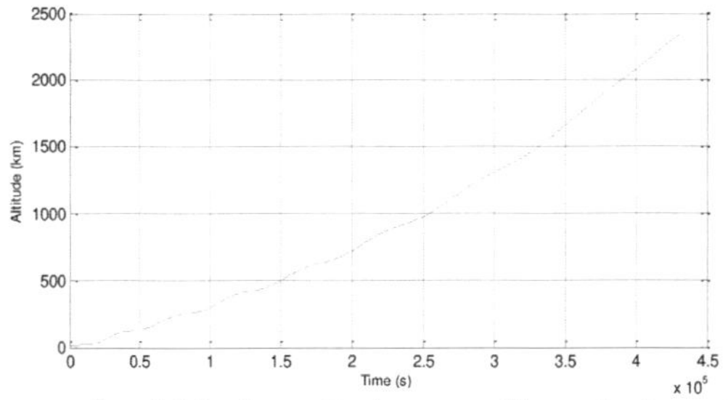

Figure 9. Altitude of the satellite during simulation 1

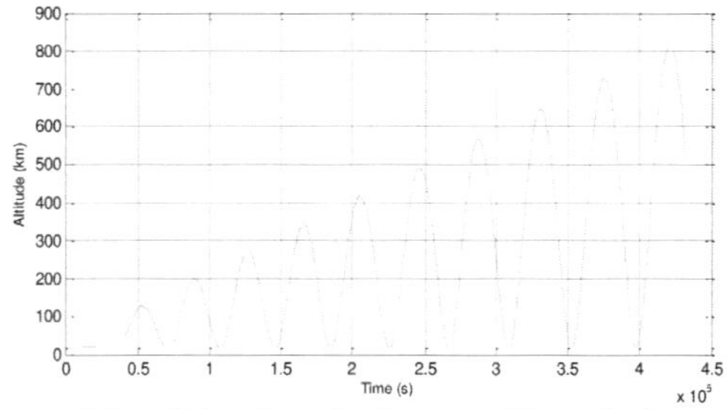

Figure 10. Altitude of the satellite during simulation 2

The Figures 11 and 12 show the disturbing force applied in the lunar satellite during the simulations. We can note that the intensity of the disturbing force decreases with the time, since the semi-major axis is increasing along the time, in other words, the distant from the lunar surface is increasing, and, therefore the disturbing due to the non-uniform distribution of mass of the Moon is becoming less relevant.

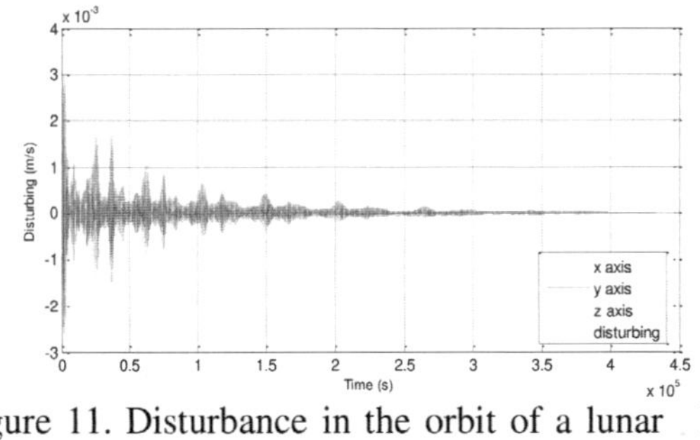

Figure 11. Disturbance in the orbit of a lunar satellite due to non-sphericity of the gravitational field during simulation 1

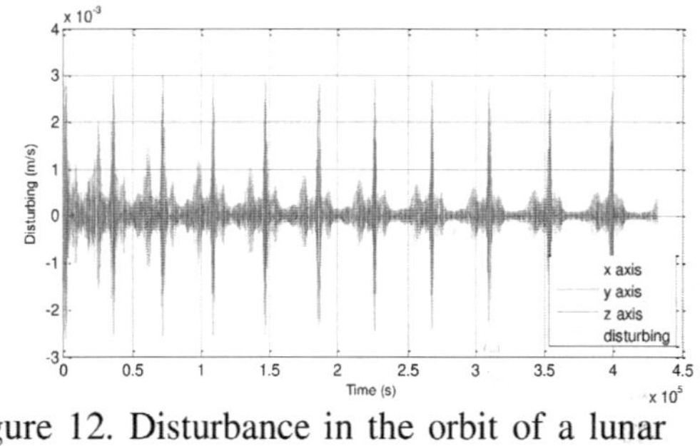

Figure 12. Disturbance in the orbit of a lunar satellite due to non-sphericity of the gravitational field during simulation 2

In the Figures 13 and 14 we can observe thrust applied during the simulations. It is possible to verify the thrust force applied in the three axes, and the control system acts separately on each axis. Note in the Figure 13 that the operation is finished when, according to the Figure 5, the semi-major axis reaches the value of 4000 km. At this point the propellant is turned off and the thrust applied tends to zero. From the Figure 14 is also possible to realize that with the passage of time the distance between two peaks or two valleys of the generated wave increases. this occurs because the semi-major axis increases during the simulation so the period of the orbit also increases.

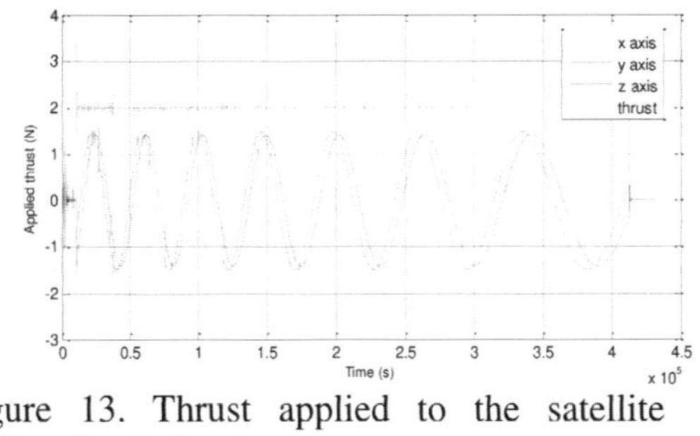

Figure 13. Thrust applied to the satellite during simulation 1

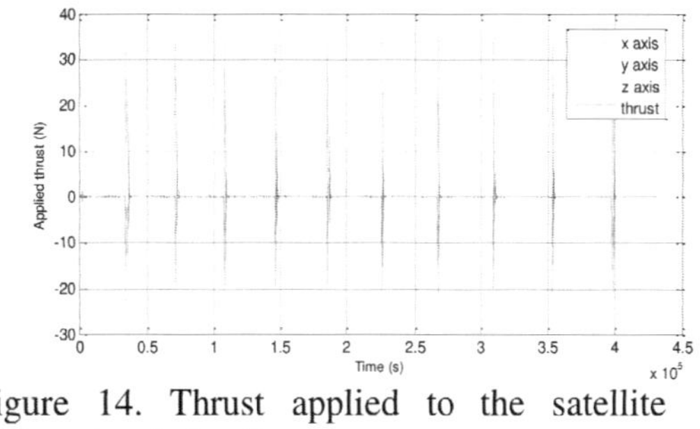

Figure 14. Thrust applied to the satellite during simulation 2

Journal of Physics: Conference Series **465** (2013) 012013 doi:10.1088/1742-6596/465/1/012013

From the Figures 15 and 16 we can analyze the fuel consumption in the simulations. We can note in the Figure 15 that at time 2000 s the propulsion system is turned on, and at the instant of 83000 s is turned off, (semi-major axis reaches 4000 km) thus the fuel consumption tends to stabilize. In the Figure 16 we can realize that each application of propulsive arc imply a significant fuel consumption, which tends to stabilize until the application of the next arc, however the consumption do not ceases to grow between arcs because the control system must act to deal with the perturbative effects that do not cease between arcs.

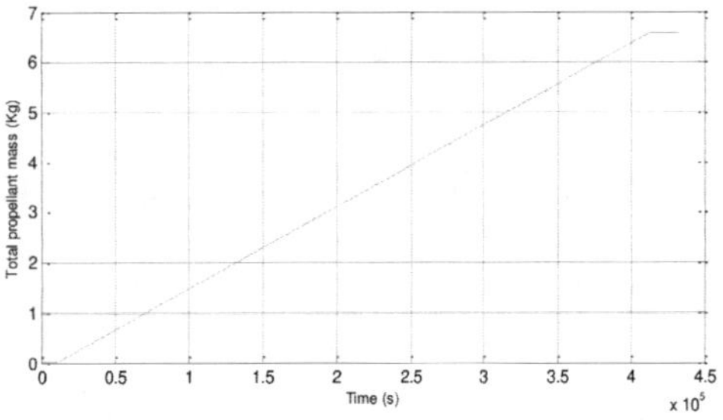

Figure 15. Mass of propellant expended during simulation 1

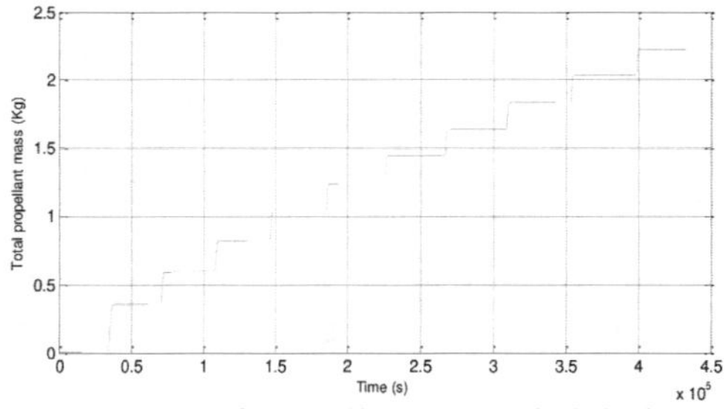

Figure 16. Mass of propellant expended during simulation 2

6. Conclusions

The results showed that the Spacecraft Trajectory Simulator, developed to analyze space missions using a closed loop control system and correct the trajectory by the application of continuous thrust is able to minimize the deviations in the path of the spacecraft when considering perturbations in the orbit due to the lunar gravitational potential of the Moon.

We can observe that the deviations in state variables values were always small, in other words, the control system was able to reduce the error in the state variables through the action of thrusters.

The Figures 11 and 12 showed that the disturbance on an artificial satellite due to the non-uniform distribution of mass of the Moon is not stable, requiring intense performance of the control system to mitigate deviations in the trajectory.

This study results are consistent with the results presented in Konopliv (2001), showing the correlation between the lunar gravitational acceleration and topography and the variation of the gravity acceleration due to non-uniform moon mass distribution. The vehicle orbital elements oscillation magnitude are in accordance with the gravity acceleration variation for the model presented Konopliv.

References

[1] KAULA, W. M. **Theory of satellite geodesy : applications of satellites to geodesy.** Waltham, MA: Blaisdell, 1966, 124 p.

[2] KONOPLIV, A. S.; ASMAR, S. W.; CARRANZA, E.; SJOGREN, W. L.; YUAN, D. N. Recent gravity models as a result of the lunar prospector mission, **Icarus**, Vol. 150, pp. 1-18, Academic Press, 2001.

[3] KUGA, H.K.; CARRARA, V.; KONDAPALLI R. R. **Satélites Artificiais – Movimento Orbital**. INPE - São José dos Campos, 2011. 111 p. Prado, A. F. B. A; Broucke, R. A. 1993. Júpiter Swing-By trajectories passing near the Earth

[4] ROCCO, E. M. Perturbed orbital motion with a PID control system for the trajectory. In: Colóquio Brasileiro de Dinâmica Orbital, 14, Águas de Lindóia,2008.

[5] TAFF, L. G., **Celestial mechanics: a computational guide for the practitioner**, New York, NY: John Wiley, 1985, 520 p.

Semi analytical study of lunar transferences using impulsive maneuvers and gravitational capture

N C Makiyama, C F de Melo, C C Celestino

Universidade Federal do ABC - UFABC, Av. dos Estados, 5001, 09210-588, Santo André, SP, Brazil

e-mail: natasha.charleaux@aluno.ufabc.edu.br

Abstract. In this work we have studied transferences between the Earth and the Moon considering bi- and tri-impulsive conventional maneuvers (like Hohmann) in order to make the acquisition of trajectories which are captured by the Moon's gravitational field through the Lagrangian equilibrium point L1. Results show that these transfer models offer reduction in the ΔV_{Total} of the mission. However, they do not require long transfer times.

Keywords: Lunar transferences, mission design, restricted three-body problem, periodic orbits and gravitational capture.

1. Introduction

In the last twenty years, the activities related with the lunar exploration were intensified by the sending of several probes [1]. In this period we can highlight the detection of water in its solid state in the lunar poles [2] and the announcement of new lunar missions with several objectives, including studies leading to the establishment of permanent bases [1, 3]. Therefore, the need for many unmanned and manned transfer missions between the Earth and the Moon can be foreseen. In this scenery, the control and the reduction of fuel consumption required for transfer maneuvers are needed in order to increase the spacecraft payload. Techniques to transfer spacecrafts between the Earth and the Moon using gravitational capture have been studied and applied since the 1980`s [4, 5]. This work presents semi analytical studies for transferences which consider the gravitational capture phenomenon and bi- and tri-impulsive conventional maneuvers that do not require long transfer times.

2. Mathematical Models

The transfer maneuvers described here are divided into three parts. In the first part, a transfer from a circular parking orbit around the Earth with altitude h_o up to a point in a capture trajectory is performed. This transference is done in two different ways: one based in a bi-impulsive Hohmann transfer and another one based in a tri-impulsive transfer [6]. The second part corresponds to the path up to the Moon by gravitational capture through the Lagrangian equilibrium point L1, and in the third part a maneuver to stabilize a spacecraft in an orbit around the Moon is performed. This latter maneuver may not be necessary for some missions. In the first and third parts the necessary ΔVs can be analytically calculated considering the dynamics of the two-body problem. In the second part, the dynamics of the Restricted Three-body Problem is considered, and the ΔVs are calculated taking into account numerical simulation results.

The Restricted Three-body Problem (R3BP) is well known in the literature [7] and the equations of motion of a particle (a spacecraft) in the xy plane of a rotating and barycentric coordinate system, also called synodic coordinate system, are given in components by

$$\ddot{x} - 2\dot{y} - x = -\left(\mu_1 \frac{x+\mu_2}{r_{13}^3} + \mu_2 \frac{x-\mu_1}{r_{23}^3} \right) \tag{1.a}$$

$$\ddot{y} + 2\dot{x} - y = -\left(\frac{\mu_1}{r_{13}^3} + \frac{\mu_2}{r_{23}^3} \right) y \tag{1.b}$$

μ_1 and μ_2 are the reduced masses of the Earth and the Moon, respectively, and their sum is 1, and the mean motion, n, of both around the common center of mass is also equal to 1. $r_{13} = [(x + \mu_2)^2 + y^2]^{1/2}$ and $r_{23} = [(x - \mu_1)^2 + y^2]^{1/2}$ are the distances between the Earth and the particle, and between the Moon and the particle, respectively. In the R3BP, the angular momentum and the energy are not conserved. However, it admits an integral of motion called Jacobi constant, C_j, given by

$$x^2 + y^2 + z^2 + 2\left(\frac{\mu_1}{r_{13}} + \frac{\mu_2}{r_{23}} \right) - \dot{x}^2 - \dot{y}^2 - \dot{z}^2 = C_j \tag{2}$$

The value of C_j defines the surfaces of zero velocity. These structures delimit the boundaries of regions in the space where the particle cannot move. The equations 1.(a) and 1.(b) have five particular solutions which correspond to five points in the synodic coordinate system in which the velocity and acceleration of a particle are null. These points are called Lagrangian equilibrium points (L1, L2, L3, L4 and L5). When $3.17216 \leq Cj < 3.18834$ a passage is opened around L1, allowing a particle to move between the Earth and the Moon. This is the case studied in this work (Fig. 1.(a)). And if $Cj < 3.17216$ another passage is opened around L2 (Fig. 1.(b)), and the particle can also move beyond the Earth-Moon system. Being C_j (L1) = 3.18834 and C_j (L2) = 3.17216 [7, 8].

The phenomena of escape and capture are temporary in the R3BP dynamics [4, 7]. Therefore, we have considered this characteristic and also a family of periodic orbits around the Moon called H2 family [8]. The initial conditions of this family are given by [6, 8]

$$(x_0; y_0; \dot{x}_0; \dot{y}_0) = (x_0; 0; 0; \dot{y}_0) \tag{3}$$

$$x_0 = \mu_1 - a_0^*(1-e_0) \quad \Rightarrow \quad x(\text{L1}) \leq x_0 < (\mu_1 - R^*_{Moon}) \tag{4}$$

$$\dot{y}_0 = -\left[\frac{\mu_2 (1+e_0)}{a_0^*(1-e_0)} \right]^{1/2} + a_0^*(1-e_0) \tag{5}$$

Where R^*_{Moon}= (R_{Moon}/384400km) is the Moon's mean radius ($R_{Moon} \approx$ 1738 km). a_0^* = (a_0/384400 km) is the semimajor axis and e_0 is the eccentricity of the particle's osculating initial orbit round the Moon.

The energy of two-bodies does not remain constant in the R3BP. However, its monitoring gives us an idea about the effects of the Earth and the Moon's gravitational fields on the motion of particles [9]. Thus, for each trajectory simulated and for each integration step, the Moon-particle two-body energy has been measured. If this energy becomes positive during the integration, the trajectory is considered an escape trajectory and it is classified taking into account the time at which the Moon-particle energy remains negative, or simply by its capture time around the Moon, and also if this escape occurs through L1 or L2.

In order to analyze the capture through the L1, we have adopted the integration for negative times. It is possible to accomplish this study for positive times, but each initial condition has a time in which it remains in orbit around the Earth, so each trajectory is captured by the Moon at different times. Because of this, the work would become extenuating.

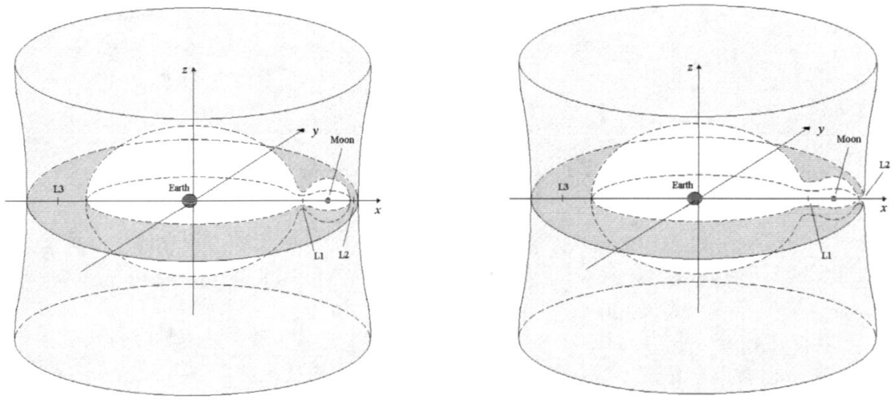

(a) (b)

Figure 1. Lagragian equilibrium points L1, L2 and L3 associated to the Earth-Moon system in the synodic coordinate system, and surfaces of zero velocities: (a) $C_j(L1) < C_j < C_j(L2)$, (b) $C_j > C_j(L2)$.

From the initial conditions defined by equations (4) and (5) it is possible to generate trajectories and to investigate them before the capture by the Moon. Figure 2 shows one of these escape trajectories with $C_j = 3.1111$. While it remains captured, it reaches 118 km above the Moon's surface. In this point, the osculating lunar orbital elements are: a_0 = 23,200 km, e_0 = 0.920, ω_0 (argument of periapsis) = 270°, Ω_0 (longitude of the ascending node) = 270° e i_0 (inclination) = 0. During its motion around the Earth, the same trajectory reaches a minimum distance of 144,795.55 km from the Earth and a maximum distance of 294,032.93 km.

3. Transfer Methods and Results

Two points can be also seen in Figure 2. They are B and B'. Point B corresponds to the point with lower distance from the Earth, and B' to the maximum distance, both belonging to the trajectory whose initial conditions of the lunar osculating orbit are a_0 = 23.200 km, e_0 = 0,920, ω_0 = 270°, Ω_0 = 270° and i_0 = 0. From an Earth circular parking orbit with a radius $R_C = h_0 + 6370$ km, two transfers are designed up to the point B and other two up to the point B'. For each point (B and B') a bi-impulsive transfer (direct transfer) and a tri-impulsive transfer (bi elliptical transfer) are designed. Therefore, four cases are considered. In the first case, a transfer ellipse tangent to the Earth orbit at points P and to the capture trajectory at point B (Fig. 3(a)) is obtained, and the application of two ΔVs is necessary. In the second case, two ellipses with the same apses line are obtained. The first one is tangent to the Earth orbit at P' and it has an apogee at point A. The second ellipse also has its apogee at point A and its perigee at point B in the capture trajectory (Fig. 3(b)). In the third case (bi-impulsive transference), a transfer ellipse tangent to the Earth`s parking orbit at a point P and also tangent to the capture trajectory at the point B' is obtained. By analogy, in the fourth case (tri-impulsive transfer), two ellipses are obtained – the first ellipse is tangent to the Earth`s parking orbit at a point P' and its apogee is at point A'. The second ellipse also has its apogee at point A' and it is tangent to the capture trajectory at point B'. In all cases, it is necessary to know the capture trajectories velocities at points B and B'. And theses quantities are known by the numerical simulations of the capture trajectories.

Figures 3(a) and 3(b) correspond to the first and second cases, respectively, and both represent the capture trajectory's acquisition at point B (the shorter distance from the Earth). The geometry transfers of the third and fourth cases are similar to the ones in these figures.

Journal of Physics: Conference Series **465** (2013) 012014 doi:10.1088/1742-6596/465/1/012014

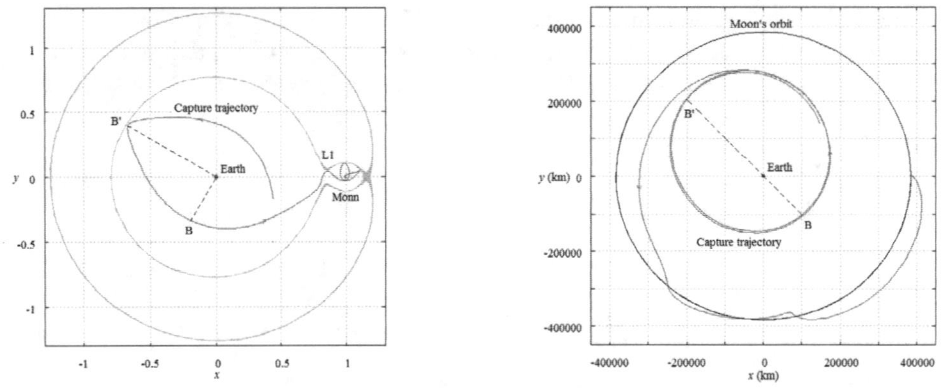

(a) (b)

Figure 2. Escape and capture trajectory by the Moon through L1, whose orbital elements of the initial osculating lunar orbit are: a_0 = 23,200 km, e_0 = 0.920, $\omega_0 = 270^o$, $\Omega_0 = 270^o$ and $i_0 = 0$.

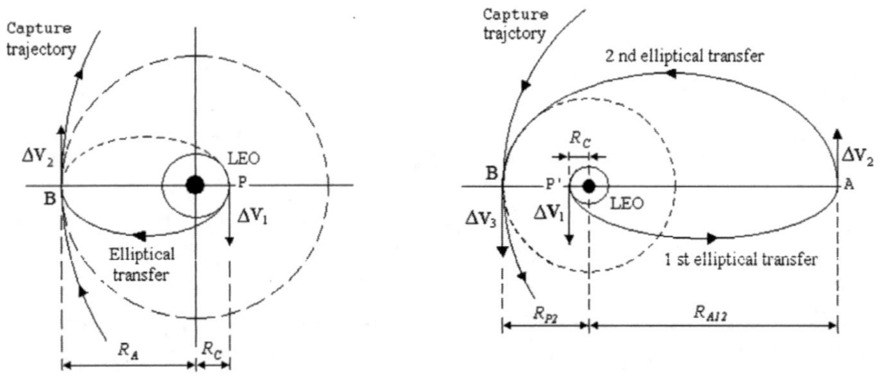

(a) (b)

Figure 3. Transfer schemes for the acquisition of the capture trajectories through the L1 at point B.

In the first and third cases (bi-impulsive transfers), ΔV_1 is analytically determined and the ΔV_2 is defined by the difference between the velocity of the capture trajectory at point B or B', and the apogee velocity of the transfer ellipse (Table 1). In second and fourth cases, that is, in the tri-impulsive transfers (bi-elliptic), ΔV_1 is also analytically determined. The determination of ellipses apogee radius, R_{A12} (Fig. 3.(b)), it is linked to the total time of transfer between the Earth and the Moon. In general, the farther from the Earth the apogee of an ellipse is, the lower its velocity. This means, in terms of orbital maneuvers, a smaller ΔV_2 to jump from the first transfer ellipse to the second one, since ΔV_2 is the difference between the apogee velocities of the two transfer ellipses (analytically calculated). Though, starting from certain values of R_{A12} the observed variation for ΔV_2 is small, however, the time of transfer becomes large. Still in the second and fourth cases, there is a ΔV_3 and it corresponds to the difference between the velocities of the capture trajectory in B or B' and the perigee of the second transfer ellipse. Figure 4.(a) shows the variation of ΔV_{Total} ($\Delta V_{\text{Total}} = \Delta V_1 + \Delta V_2 + \Delta V_3$) for one bi-elliptical transfer as a function of the apogee distance from the Earth, R_{A12}, and in Figure 4.(b), ΔV_{Total} is given as function of the time of transfer. The values are obtained for the trajectory shown in Figure 2 considering that the acquisition of the capture trajectory occurs at point B'. For times of transfers larger than 120 days, which corresponds to $R_{A12} > 1.5\text{x}10^6$ km, ΔV_{Total} is at 3.785 km/s, a value having little variation. Thus, starting from a certain point, the time of transfer becomes more important in order to determine the transfer parameters than ΔV_{Total}.

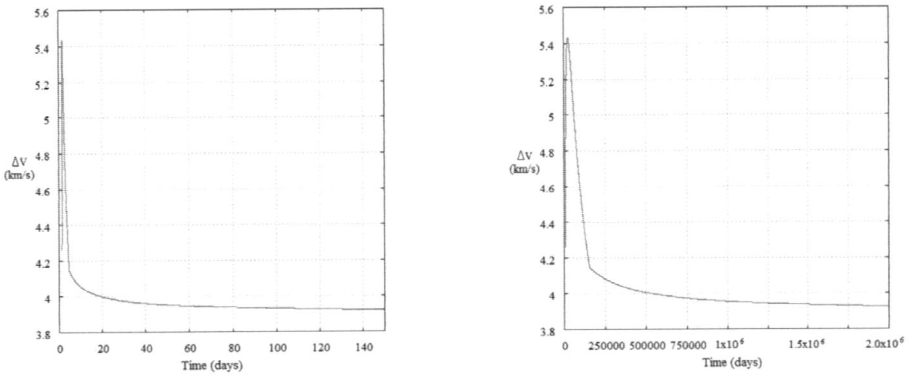

(a) (b)

Figure 4. (a) ΔV_{Total} ($= \Delta V_1 + \Delta V_2 + \Delta V_3$) for a bi-elliptical transfer among P' and B' as function of the apogee distance from the Earth, R_{A12}. (b) V_{Total} among P' and B' as function of the time of transfer.

Tables 1 and 2 show the results for the capture trajectory in Figure 2. But the technique for calculating ΔV_{Total} can be considered for any trajectories whose initial conditions satisfy eq. (4) and (5). Table 1 brings the values for the bi-impulsive transfer for cases 1 and 3 that correspond to a direct transfer between points P and B, and P and B'. Table 2 brings some values for the tri-impulsive transfer (bi-elliptical) for ΔV_{Total} and makes the relationship between them and the time of transfer, and distance from the apogee to the Earth (R_{A12}) for case 4. It is possible to observe that for ΔV_{Total} = 3.800 km/s, the time of transfer is 82 days, and for ΔV_{Total} = 3.785 km/s, 123 days are needed to complete the maneuver. In other words, there is an increase of 41 days for a reduction of only 15 m/s in the ΔV_{Total}. For a ΔV_{Total} = 3.900 km/s, the time of transfer is of 33.8 days. This value is practically similar to the one in the third case (direct transfer among the points P' and B). An analysis of the second case is similar to the fourth case, however, ΔV_{Total} below 4.000 km/s is obtained only for time of transfers longer than 100 days. This makes any application of this maneuver unfeasible, since there are faster and more economical alternatives.

Finally, it remains to analyze the one which will happen to the spacecraft after it is captured by the Moon. Naturally, it will remain in an orbit around the Moon for some time until a new escape happens. This time may be enough for the development of a lot of missions. The trajectory shown in Figure 2, for instance, this time is 37.4 days. But, if this time is not enough to accomplish a mission. The spacecraft can be stabilized in a keplerian orbit around the Moon. To stabilize the spacecraft in an orbit around the Moon, it would be necessary to apply an extra ΔV in a point that corresponds to the larger distance from the Moon (in the Moon's sphere of influence), or in the point of smaller distance, both extra ΔV are analytically calculated. If this maneuver is necessary, the extra ΔV should be added to the ΔV_{Total} of tables 1 and 2.

4. Conclusions

The transfers presented in this work are planed starting from the two- and three-body problem dynamics, and they correspond to alternative ways to study Earth-Moon transfers, and they are perfectly feasible. The results, analyzed in terms of ΔV_{Total} and the time of transfer (here presented for the purpose of an example) are attractive; especially if they are compared to the Patched-conic approximation [10] used in the Apollo missions, for instance, whose ΔV_{Total} ranged between 4.100 and 4.500 km/s. The choice, or study, of these alternative transfers necessarily involves the analysis of the time of transfers. In the cases of directed bi-impulsive transfers, for example, between points P and B', they present ΔV_{Total} = 3.900 km/s and the time of transfer is not very long. In this case, 33.26 days. This is a long time for manned missions, but taking into account unmanned missions of many natures, as supply transport, given the ΔV_{Total} reduction; these transferences represent consistent alternative.

Journal of Physics: Conference Series **465** (2013) 012014 doi:10.1088/1742-6596/465/1/012014

Table 1. First and third cases, bi-impulsive transfers between P and B and P and B' for the trajectory of Figure 2.

Earth parking orbit and transfer ellipse variables		Case 1	Case 3
Radius of Earth parking orbit	$R_c = h_0 + 6{,}370$ km, ($h_0 = 200$ km)	6,570 km	6,570 km
Velocity of the Earth parking orbit	$V_c = [GM_E/R_c]^{1/2}$	7.788 km/s	7.788 km/s
Distance Earth of points B and B'	R_A	144,795.55 km	294,032.93 km
Semimajor axis	$a = (Rc+R_A)/2$	75,682.17 km	150,301.5 km
Eccentricity	$e = 1 - R_c/a$	0.9131	0.9562
Perigee velocity	$V_p = [GM_E(1/a + 2/R_c)]^{1/2}$	10.772km/s	10.893 km/s
Apogee velocity	$V_A = [GM_E(1/a + 2/R_A)]^{1/2}$	0.489 km/s	0.244 km/s
ΔV_1	$\Delta V_1 = V_p - V_c$	2.984 km/s	3.104 km/s
ΔV_2	$\Delta V_2 = \|V_A - V_B\|$ or $\|V_A - V_{B'}\|$	1.431 km/s	0.696 km/s
ΔV_{Total}	$\Delta V_{Total} = \Delta V_1 + \Delta V_2$	4.415 km/s	3.900 km/s
Time of transfer	$Tof = 2\pi(a^3/GM_T)^{1/2}$	1.2 days	3.66 days
Time of periselenium trajectory	T_1 numerically calculeted	23.8 days	29.6 days
Total time of transfer	$T_{total} = Tof + T_1$	25 days	33.26 days

M_E (= 5.9742x10^{24} kg) is the Earth's mass, G (= 6.67x10^{-20} km^{3}s^{-2}/kg) is the Gravitational constant, V_B (= 1.920 km/s) is the capture trajectory velocity at the points B and $V_{B'}$ (= 0.940 km/s) at the point B'.

Table 2. Case 4: tri-impulsive (bi elliptical between P' and B').

Apogee distance (km)	ΔV_{Total} (km/s)	Total time of transfer (days)
1,743,349	3.785	123.78
1,193,109	3.800	82.07
314,205	3.900	33.83

References

[1] Foing, B.H.; Ehrenfreund, P. 2008 Jurney to the Moon: Recent results, science, future robotic and human exploration. Advances in Space Research, vol. 42, p 235-237.

[2] Regeon, P.A.; Lynn, P.R.; Johnson, M.; Chapman, R.J. 1996 The Clementine Lunar Orbiter, In: 20th International Symposium on Space Technology and Science, Gigu, Japan 19-15, May 1996. Proceedings of 20th ISSTS.

[3] NASA'S Exploration Systems Architeture Study – ESAS / 2005. www.nasa.gov/pdf/140631main_ESAS.

[4] Belbruno, E.A.; Miller, J.K. 1990 A Ballistic Lunar Capture Trajectory for Japanese Spacecraft Hiten. Pasadena, CA, Jet Propulsion Laboratory – California Institute of Technology, JPL IOM-90.4-1731.

[5] Foing, B.H., Racca, G. D., Marini, A., Evrard, E., Stagnaro, L, Almeida, M., Koscny, D., et al. 2006 SMART-1 mission to the Moon: Status, first Results and Goals. Advances in Space Research, vol. 37, p 6-13.

[6] de Melo, C.F.; Macau, E.E.N.; Winter, O.C. 2007 Numerical Study about Natural Escape and Capture Routes by the Moon via Lagrangian Points L1 and L2. Advances in Space Research. vol. 40, p 83-95.

[7] Murray, C.D. and Dermott, S.F. 1999 *Solar System Dynamics,* UK: Cambridge University Press.

[8] Broucke, R.A.: 1968, *Periodic Orbits in the Restricted Three-Body Problem with Earth-Moon Masses*, JPL Technical Report 32-1168.

[9] Winter, O.C., Vieria Neto, 2002 E. Distant Stable Direct Orbits Around the Moon, A&A 393, 661-671.

[10] Bate, R.R.; Mueller, D.D.; White, J.E. Fundamentals of Astrodynamics. USA: Dover Publications Inc., 1971. 455p.

A study of low velocities regions near the Roche lobe during the gas giant planets formation

Ricardo A. Moraes
UNESP - Universidade Estadual Paulista, Guaratinguetá, CEP 12516-110, SP, Brazil
E-mail: ricardo.moraes07@gmail.com

Ernesto Vieira Neto
UNESP - Universidade Estadual Paulista, Guaratinguetá, CEP 12516-110, SP, Brazil
E-mail: ernesto@feg.unesp.br

Abstract. In this paper were studied regions close to the Roche lobe of a planet like Jupiter, in order to find regions with low velocities. We simulated a two dimensional and non-self-gravitating disk, where tidal and viscous torques are considered, using the hydrodynamic numerical integrator FARGO 2D. As stated earlier we are interested in find low velocities regions for in future works study the possibility of satellites formation in these regions.

1. Introduction

It's believed that satellites of a giant gaseous planet were formed soon after the own planets, from material which was not accreted on the planet in its last formation phase. Our work analyzes some special regions near the Roche lobe of giant planets, regions where the gas has low velocities. We believe that these regions will occur accretive collisions between the material not accreted by planet and than, due to low velocity, a body could be formed. According to [6] the idea is that if the accreting inflow has sufficiently low angular momentum per unit mass, then the satellites formed with a circumplanetary disk will occupy a small region within the Hill sphere, such arguments justify our interest in the study these regions. Thus, we simulated the formation of a planet like Jupiter from a typical protoplanetary disk, using the core accretion model described in [1].

This work is structured as follows, in the section 2 we explain how a giant gaseous planet form according with the core accretion model. In the section 3 we describe the numerical procedures used in our simulation. In the section 4 we discuss about our results, in the section 5 we present the conclusions about the work, and in the section 6 comments about the next works that will arise from the results provided in this work.

2. Formation of Giant Gaseous Planets

Two teorical model are are frequently used to explain the formation of the giant gaseous planets, although they are not the only ones, the disk instability model and the core accretion model. From [1], the disk instability model for giant planet formation is predicated on the assumption that the gaseous protoplanetary disk was massive enough to be unstable in such a way that instabilities arise from its own self-gravity, and that the outcome of these instabilities is the fragmentation of the disk into massive planets.

The core accretion model is the most acceptable model to explain the formation of all planets, including the giant ones, of our Solar System [9,10], we adopted that model to study the planetary formation in our simulations.

The core accretion model can be separeted in four stages, as showed in the figure 1. The formation of the solid core; hydrostatic growth; the runaway growth and the phase of ending accretion. In the first stage, the solid protoplanetary disk grows via a sucession of two-body collisions until it becomes massive enough to retain a significant gaseous atmosphere or envelope [1]. The process of formation of the planetary embryo requires, initially, low velocities in the collisions between the planetesimals, that are basically formed by rock and ice, on the feeding zone, assuming that the collisions between the planetesimals are totally accretive, that means, all the physical collisions are completely inelastic [5] we have that the initial solid core of the planet starts to form in this stage. That is considered a quick event, around 0.5 Myr after the formation of the star, and is similar to the formation of the terrestrial planets. The rate of planetesimals accretion can be calculated by [5]:

$$\dot{M}_{core} = \pi R^2 \sigma_{core} \Omega F_g \tag{1}$$

where R is the radius of the accreting body, σ_{core} is the surface density of planetesimals in the solar nebula, Ω is the orbital frequency and F_g is the gravitational enhancement factor.

In the second stage the protoplanet and the gaseous envelope are both in hydrostatic equilibrium, this equilibrium is broke when the protoplanet exceeds a critical mass due the constant accretion of material. The critical mass isn't a constant, it's a function of planetesimal accretion rate and opacity in the gaseous envelope, the estimates of the critical mass required range from $5M_\oplus$ [2] to $20M_\oplus$ or more [10]. From [1], the envelope opacities $\kappa_R \geq 10^{-2} cm^2 g^{-1}$ the critical core mass can be approximated by:

$$M_{crit} \approx 7 \left(\frac{\dot{M}_{core}}{10^{-7} M_\oplus yr^{-1}} \right)^q \left(\frac{\kappa_R}{1 cm^2 g^{-1}} \right)^s M_\oplus \tag{2}$$

where q and s are both estimated to lie in the range 0.2-0.3

Once this mass is exceeded, the protoplanet enters in a stage called runaway growth, in this phase of the formation process there is no hydrostatic equilibrium, in other words, all material accumulated by the planet during the previous steps collapses to its core. For massive planets the most of its planetary envelope is accreted during this stage, which is a quick process, about 10^5 years.

In the final stage of the formation, the protoplanet do not accretes gas anymore, either as a consequence of the planet opening a local gap in the disk or as a consequence of the dissipation of the entire protoplanetary disk [1]. The energy released by accretion increases considerably the temperature of the planet and of its gaseous envelope, as a consequence, the icy planetesimals which were accreted dissolves causing the core to be composed only by the rocky material, ending the whole process of formation of a gas giant planet.

3. Numerical procedures

For the development of the numerical calculations we used the hydrodynamic numerical integrator FARGO 2D [8] and we set up our simulations in the conditions used in [7].

We simulated a two-dimensional protoplanetary disk, considering the tidal and viscous torques, with neglected self-gravity. A typical protoplanetary disk has a thickness of radius given by the ratio $H/r \approx 0.05 - 0.1$ [7], we adopted a thickness to radius of the disk as $H/r = 0.05$ and non reflecting inner boundary conditions [4].

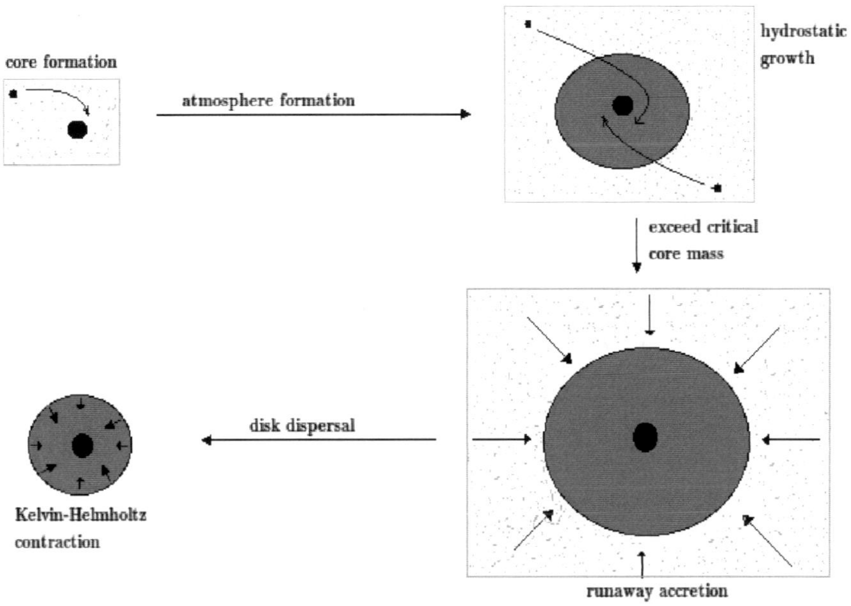

Figure 1. Scheme of formation of a giant planet via core accretion. Adapted from [1].

In our model we adopted the viscous forces per unit area of Navier-Stokes equations for modeling the viscous forces. The origin of the coordinate system is taken to be in the center of the star, and we used cilindrical coordinates (r, θ). Thus, the equations of motion are given by [7]:

$$\frac{\partial \Sigma}{\partial t} + \nabla (\Sigma \mathbf{u}) = 0 \tag{3}$$

$$\frac{\partial p_r}{\partial t} + \nabla (p_r \mathbf{u}) = \Sigma r \left(\frac{u_\theta}{r} + \Omega_p \right)^2 - \frac{\partial p}{\partial r} - \Sigma \frac{\partial \Phi}{\partial r} + f_r \tag{4}$$

$$\frac{\partial j}{\partial t} + \nabla (j \mathbf{u}) = -\frac{\partial p}{\partial \theta} - \Sigma \frac{\partial \Phi}{\partial \theta} + f_\theta \tag{5}$$

where Σ is the surface density of the disk, $\mathbf{u}$ is the flow velocities, in terms of its components, can be written by $\mathbf{u} = u_r \hat{r} + u_\theta \hat{\theta}$, $p_r = \Sigma u_r$ is the radial momentum per unit area, p is the vertically integrated gas pressure, Ω_p is the angular velocity of the planet, $j = \Sigma r (u_\theta + \Omega_p r)$ is the gas angular momentum per unit area, $\mathbf{f} = f_r \hat{r} + f_\theta \hat{\theta}$ is the viscous force per unit area and Φ is the gravitational potential smoothed in the neighborhood of the planet, given by [7]:

$$\Phi = -\frac{G M_s}{r} - \frac{G M_p}{\left(|\mathbf{r} - \mathbf{r}_p|^2 + r_{sm}^2 \right)^{\frac{1}{2}}} \tag{6}$$

where M_s is the mass of the star, M_p is the mass of the planet, $\mathbf{r}_p$ is the planet position and r_{sm} is the smoothing length of the planet's potential.

For the value of the kinematic turbulent viscosity we have, $\nu = \mu / \Sigma$, where ν can be expressed in terms of the α prescription given in [11] as,

$$\nu = \alpha c_s H \tag{7}$$

where c_s is the sound speed, H is the thickness of the protostellar disk and the parameter α is dimensionless.

Journal of Physics: Conference Series **465** (2013) 012015 doi:10.1088/1742-6596/465/1/012015

4. Results

In this section we will discuss about the results obtained in our simulations using the numerical procedures described in the last section.

We work with an accretive disk, the planet, that have caracteristics of a Jupiter and was fixed in $r = 1$ and $\theta = \pi$, or correspondingly $x = 1$ and $y = 0$ in cartesian coordinates, in a grid of 252 by 384, a zoom was made in the region around the Roche lobe, the disk has an inner radius of 0.4 and a outer radius of 2.5, we simulated 140720 Jupiter orbits.

The figure 2.a shows the gas density in the simulated disk and inside the circle we have around the region of Roche lobe, in this figure we can see the gap opened by the planet.

The figure 2.b shows the region inside the circle of figure 2.a. We can clearly see the low density in two regions close to the planet, what can be an evidence of regions with low velocities that we are looking for, we can't conclude this looking to this figure, for this we present the figure 3.

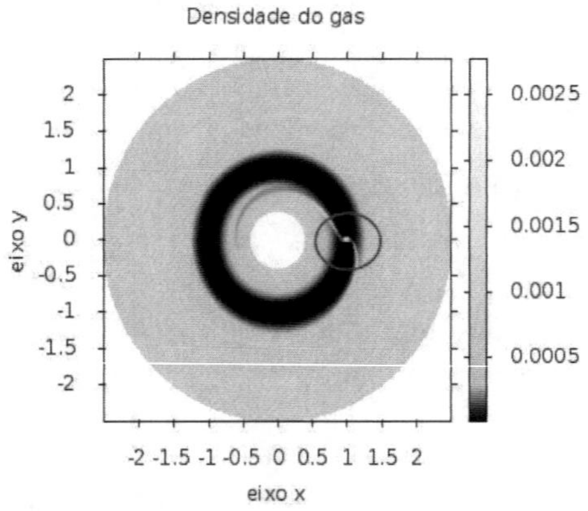

(a) Gas density in all region around the planet.

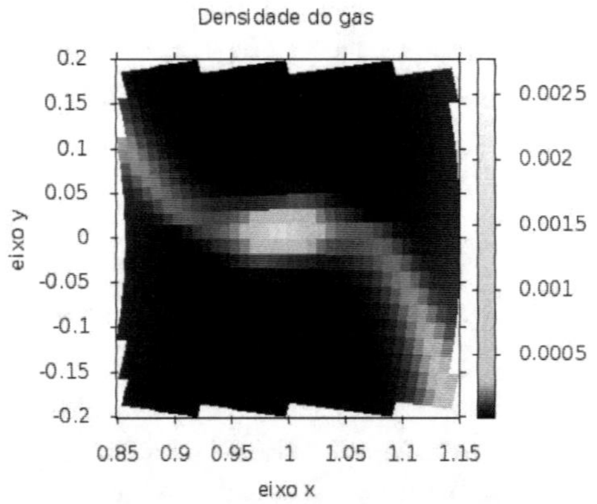

(b) Zoom aplied in the figure 2.a.

Figure 2. Gas density around the planet.

In figure 3 we show the velocity field around the simulated planet, this figure assists our analises about the velocities close to the planet. We see in this figure, that near the planet the velocities converges to the planet creating a vortex, we believe that this vortex could be responsable for the accretion of planetesimals. We also identified regions of escape of planetesimals, these regions have high velocities and pointing outward the center of the planet. There is also two peculiar regions in this figure, it is slightly above and below the vortex of the planet, in these regions we can observe that the velocities field is to weak, these regions are, indeed, the regions that we are looking for, which we called low velocities regions.

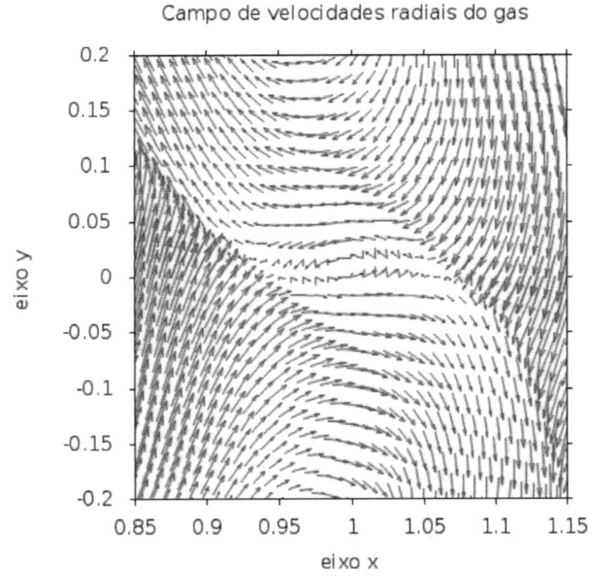

Figure 3. Velocities field around the planet.

5. Conclusions

With the simulations described in the previous sections we found the low velocities regions close to the Roche lobe of giant gaseous planet, as showed in figure 3. The results found in this work agrees with the results presented by [7] and [4], who analyzed these regions before. Besidesfinding the low velocities regions, we also obtained results that given us an estimation of the location of these regions

6. Future works

The results presented in this paper are preliminary results, our goal was show that there are regions with low velocities close to the Roche lobe of giant gasous planet. With this result, we will study the possibility of the satellite formation in these regions, the goal of the new project is to show that regions with low velocities close to the Roche lobe can be good sites to satellite formation, to do that, we must simulate planetesimals near to the planet, in the region of interest.

References

[1] Armitage, P. J., 2010, *Astrophysics of Planet Formation* (Cambridge University Press, Cambridge)
[2] Hubickyj, O., Bodenheimer, P., & Lissauer, J. J. 2005 *Icarus* **179** 254

[3] Irwin, P. G. J., 2006, *Giant Planets of Our Solar System* (Praxis Publishing Ltd, Chichester)
[4] Kley, W. 1999, *MNRAS* **303** 696
[5] Lissauer, J. J. & Stevenson, D. J. 2007, *Protostars and Planets* V, B. Reipurth, D. Jewitt, and K. Keil(eds.) (University of Arizona Press, Tucson) p.591
[6] Lubow, S. H., Martin, R G., 2013, *MNRAS* **428** 2668
[7] Lubow, S. H., Seibert, M., Artymowicz, P., 1999, *ApJ* **526** 10001
[8] Masset, F. 2000, *Astronomy & Astrophysics* **141** 165
[9] Mizuno, H. 1980, *Progress of Theoretical Physics* **64** 544
[10] Pollack, J. B., Hubickyj, O., Bodenheimer, P., *et al.* 1996, *Icarus* **124** 62
[11] Shakura, N. I., Sunyaev, R. A., 1973, *Astronomy & Astrophysics* **24** 337

Elements of Non-Linear Hamiltonian Dynamics

R Egydio de Carvalho

Universidade Estadual Paulista-UNESP, 13506-900 Rio Claro, SP–Brazil

e-mail: regydio@rc.unesp.br

Abstract. In this paper we present a set of generic results on Hamiltonian non-linear dynamics. We show the necessary conditions for a Hamiltonian system to present a non-twist scenario and from that we introduce the isochronous resonances. The generality of these resonances is shown from the Hamiltonian given by the Birkhof-Gustavson normal form, which can be considered a toy model, and from an optic system governed by the non-linear map of the annular billiard. We also define a special kind of transport barrier called robust torus. The meanders and shearless curves are also presented and we show the most robust shearless barrier associated with the rotation numbers.

Keywords: isochronous resonances, robust torus, meandering tori, shearless curves.

1. Introduction

Homoclinic chaos, as is known chaos in Hamiltonian systems, is theoretically supported by the as well-known KAM theory [1] and by the Poincaré-Birkhoff theorem [2]. Roughly speaking, KAM theory establishes that a dynamical system of two degrees of freedom governed by a Hamiltonian of the form,

$$H(J_k,\theta_k) = H_0(J_k) + \varepsilon H_1(J_k,\theta_k) \quad ; \quad k=1 \text{ and } 2 \tag{1}$$

which is integrable for $\epsilon = 0$, still has many invariant tori even if $\epsilon \neq 0$, but ϵ small enough.

The simplest version of the theorem associated with the KAM theory, implies that the Hamiltonian must be sufficiently smooth, and the unperturbed Hamiltonian H_0 to be nonsingular, this is it must satisfy the non-degeneracy, or twist, condition [3], $\det\left|\frac{\partial \dot{\theta}_k}{\partial J_j}\right| \neq 0$, where $\dot{\theta}_k = \frac{\partial H_0(\underline{J})}{\partial J_k} = \omega_k$, are the unperturbed frequencies. It means that the rotation number $\rho = \frac{\omega_1}{\omega_2}$ is a monotonically increasing, or decreasing, function. If ρ is a rational number, it defines a rational or resonant torus and similarly it defines an irrational torus when ρ is irrational. When the perturbation is turned on it acts on a resonant torus and as it becomes stronger the neighbourhood of this torus is also affected defining a resonance region. Typically, this resonance region is constituted by island chains formed by an even number of periodic orbits, which has been continued from the unperturbed system, libration and rotation tori, deformed irrational tori and chaotic orbits. These periodic orbits have alternating stability and the corresponding island chains are called Poincaré–Birkhoff chains.

However, if the rotation number presents an extreme, the non-twist scenario is established and we observe the birth of isochronous resonances [4], which correspond to simultaneous resonances of same period, and new phenomena, not foreseen by the Poincaré's theorem, appear. A new one very important is the occurrence of heteroclinic reconnections among manifolds of hyperbolic fixed points of different isochronous resonances [5]. The new structures that locally appear in the phase space are associated with the possible values of the non-monotonic behaviour of the rotation number. It is important to point out that there is an extension of the KAM theorem, for discrete equations, for non-twist dynamics [6].

2. Resonant Normal Form and the Isochronous Resonances

In order to introduce the isochronous resonances we will consider the toy model, Resonant Normal Form (RNF) [7,8], for an autonomous system with two degrees of freedom described in angle-action variables (J,θ). The RNF is a Hamiltonian function that consists of an expansion series in all variables around an elliptic fixed point. The terms that do not contain the variables angle are grouped and define what we call unperturbed Hamiltonian H_0 (J_1,J_2). The terms containing only the angle θ_1, besides the actions, define the first perturbation H_1 (J_1,J_2,θ_1), which is an integrable perturbation. The non-integrability of the system is only obtained when both angles appear in the total Hamiltonian. So, the Hamiltonian function can be given by,

$$H(J_1,J_2,\theta_1) = \sum_{k=0}^{n} a_k\, J^k + \varepsilon\, f(J_1,J_2)\cos(r\,\theta_1) \quad , \text{ with } \; a_0 = J_2 \tag{2}$$

where $H_0(J_1,J_2) = \sum_{k=0}^{n} a_k\, J^k$ and $f(J_1,J_2)$ is an analytic and continuous function. The resonance condition is:

$$\left(\dot{\theta}_1\right)_0 = \text{a polynomial of order (n} - 1\text{) in } J_1 = \frac{\partial H_0}{\partial J_1} = 0 \tag{3}$$

and defines (n-1) Isochronous Resonances, each one at each root of this polynomial. The equations of motion are

$$\dot{J}_1 = r\,\varepsilon\, f(J_1,J_2)\,\sin(r\theta_1) \quad , \quad \dot{\theta}_1 = H_0'(J_1,J_2) + \varepsilon\, f'(J_1,J_2)\cos(r\theta_1) \tag{4}$$

and for,

$$\theta_1 = \frac{k\,\pi}{r} \;\Rightarrow\; k = 0, \ldots, 2r-1 \tag{5}$$

we have 2r fixed points located at values of J_1 satisfying

$$H_0'(J_1,J_2) + (-1)^k\, \varepsilon\, f'(J_1,J_2) = 0 \tag{6}$$

The stability distribution of fixed points, in general, depends on the function f (J). If J_i and J_{i+1} are two consecutive roots of the equation above, then if,

f (Ji) f (J_{i+1}) > 0 ➔ defocalized chains ➔ type 1, f (Ji) f (J_{i+1}) < 0 ➔ focalized chains ➔ type 2, and the corresponding scenarios can be visualized according to picture of Figure (1) [9]. Figure (1) is a permitted reproduction of the original work: Voyatzis G, Meletlidou E and Ichtiaroglou S 2002 Chaos, Solitons & Fractals **14** 1179, authorized by Elsevier through the License Number: 3124200437200.

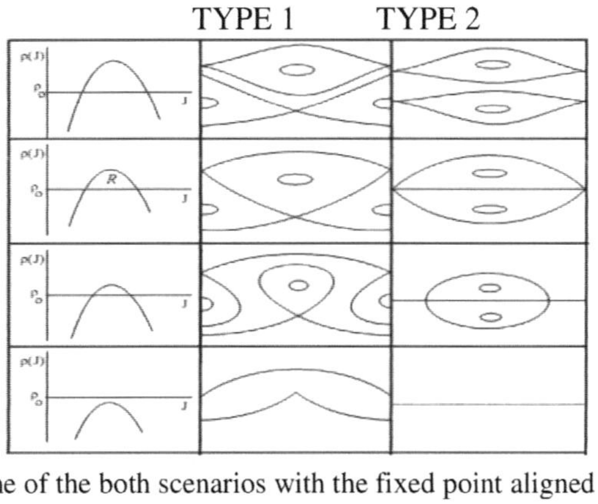

Figure 1: Scheme of the both scenarios with the fixed point aligned or defocused.

In order to present a numerical illustration of the occurrence of isochronous resonances, we have adjusted as the unperturbed Hamiltonian H_0 as the function f (J_1,J_2) and we got,

$$\mathrm{H}=\left\{\left[J_2-\frac{a}{2}(4J_1-c)^2+\frac{1}{4}(4J_1-c)^4\right]+\alpha\left\{\left[b(4J_1-c)^2-a\right](4J_1{}^2)(J_2-J_1)^{1/2}\cos(\theta_1)\right\}\right\} \tag{7}$$

where a, b, c, α are adjustable parameters and the action J_2 is a constant of motion since θ_2 is a cyclic variable. Since H_0 is of fourth order in J_1, from equation (2) we identify the position of the three isochronous resonances in the phase space (J_1,θ_1), their positions are

$$(J_1)_+=\frac{c+\sqrt{a}}{4} \quad , \quad (J_1)_m=\frac{c}{4} \quad , \quad (J_1)_-=\frac{c-\sqrt{a}}{4} \tag{8}$$

where "m" stands for the middle resonance, while "+" the upper resonance and "-" the lower one. In the plots of figure (2) below we observe the three island chains of isochronous resonances for four different values of the perturbation parameter α.

The appearance of resonance chains of same order is the signature of non-twist systems. Besides this effect, the integrable scenario propitiated by the Hamiltonian (7) allows us to observe in a clearer way, in figure (2), the mechanism of the overlap of resonances [10]. As we increase α, the separatrix move toward each other and the overlap starts with a topological rearrangement of the resonance structures. It is worth mentioning that outside this region of isochronous resonances the system behaves as a twist one. The plots of figure (2) are permitted reproduction of the original work: Egydio de Carvalho R, Martins C G L and Favaro G M 2009 Braz. J. Phys.**39** 606, authorized by the Brazilian Physics Society.

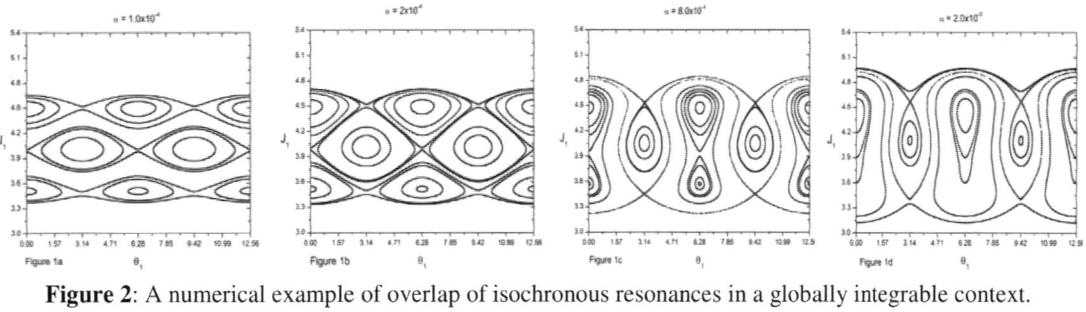

Figure 2: A numerical example of overlap of isochronous resonances in a globally integrable context.

Journal of Physics: Conference Series **465** (2013) 012016 doi:10.1088/1742-6596/465/1/012016

3. The dynamical barrier robust torus

If the function f (J_1,J_2) of the Hamiltonian (2) is a polynomial in J_1 with real roots, then in each root, the perturbation will be algebraically zero. So, invariant curves of H_0 will continue in the perturbed system and we call them by Robust Tori (RT) [3,11].

In the Hamiltonian (7) we identify this function f (J_1,J_2) as,

$$f(J_1,J_2)=\left[b(4J_1-c)^2-a\right](4J_1^{\,2})(J_2-J_1)^{1/2} \tag{9}$$

This pre-factor may be null in three different cases: i) $J_1 = 0$, which corresponds to the elliptic point of the system around of which the normal form has been expended and do not add any new information; ii) $J_2 = J_1$, but this case is avoided because numerically we take $J_2 >> J_1$; and iii) $J_1 = \frac{c \pm \sqrt{a}}{4}$, which defines two RT in the phase space of the total Hamiltonian.

If the RT are intercalated with the isochronous resonance chains then we obtain the scenario type 2 cited above. On the other hand, if they are external to the chains, we obtain the scenario type 1. Both cases can be seen in figure (3) using the Hamiltonian (7) [11]. The plots of figure (3) are permitted reproduction of the original work: Egydio de Carvalho R and Favaro G M 2005 **350** 173, authorized by Elsevier through the License Number: 3172610159127.

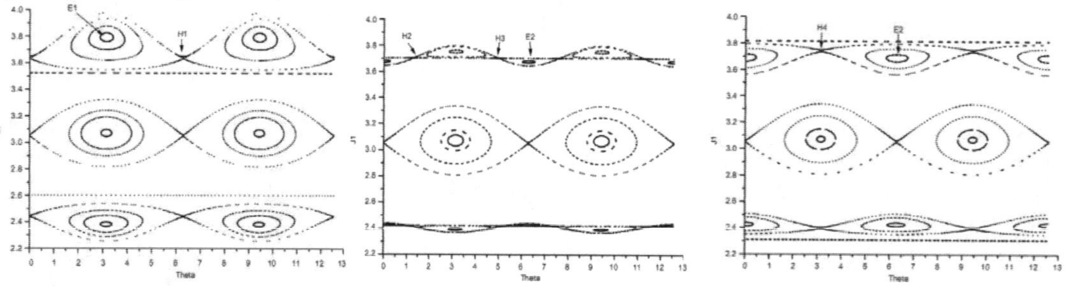

Figure 3: Interaction between the robust tori with the aligned island chains and the defocused scenario.

The movement of the RT from the intercalated to outside configurations is controlled by the parameter b and the mechanism that drives this transition is the successive bifurcation that occur when the RT interact with the separatrix of the islands. The letters H and E in figure (3) corresponds to hyperbolic and elliptic fixed points respectively. A remarkable effect of the presence of the RT is the transition from the focused scenario, where we can observe an alignment of the fixed points of same stability, to defocused scenario or vice-versa.

Since the RT can survive to any generic perturbation that has a polynomial pre-factor, with real roots then, in these situations, they constitute dynamical barriers that separate the phase space in different regions. The existence of this kind of transport barrier can play an important role in different branches of the science, particularly to confinement of plasma in tokamaks [12].

4. Meanders and shearless curves

The isochronous resonances can interact and go into overlapping producing a local, but wide, topological rearrangement. After the overlap, a region with meandering tori appears in the phase space. This kind of tori also corresponds to transport barriers for the dynamics in the phase space. They behave differently under the action of a generic perturbation and their robustness is associated with their rotation number. The destruction of each meandering torus transforms its neighbourhood in a stickiness beach and the last meandering torus to be destroyed is called shearless torus [13,14]. The authors of [14] have shown numerically that other overlap can locally occur inside each meandering region and they called this kind of overlap as second-order reconnection. Other reconnections can happen in smaller scales, inside the meander structures, and thus the order of the reconnections can also increases. After the destruction of the shearless torus a strong stickiness zone emerges. This region still plays an important role to the dynamics, because it defines local transport barriers [14,15,16,17].

We illustrate the appearance and the robustness of the meandering tori from the labyrinthic standard map non-twist map introduced in [18],

$$x_{n+1} = x_n + a\,(1 - y_{n+1}{}^2) \quad , \quad y_{n+1} = y_n - b\{\,\sin(2\pi\, x_n) + \sin(\eta\, 2\pi\, x_n)\,\} \tag{10}$$

where a, b and η are adjustable parameters. A remarkable effect is produced through the variation of η. If η = 0 or 1, the system present two isochronous island chains with a pair of fixed points, one being elliptic and the other hyperbolic. Nevertheless, for any other integer η > 1 there occur bifurcations of these fixed points and will exist so many pairs of fixed points as the value of η in each island chain. In figure (4) we present the case for η = 4, in which we can observe four elliptic fixed points as well the four hyperbolic ones in both chains. In all plots from this map we used a = 1.006. The pink lines correspond to a region with some meandering tori. The plot in the right is the rotation number of the invariant curves and the peak, with a pink arrow, corresponds to an extreme of the rotation number, which is associated with a meandering torus of the pink region. Figures (4,5,6) are authorized reproductions:"© IOP Publishing. Reproduced by permission of IOP Publishing. All rights reserved", of the original work: Martins C G L, Egydio de Carvalho R, Caldas I L and Roberto M 2011 J. Phys. A **44** 045102.

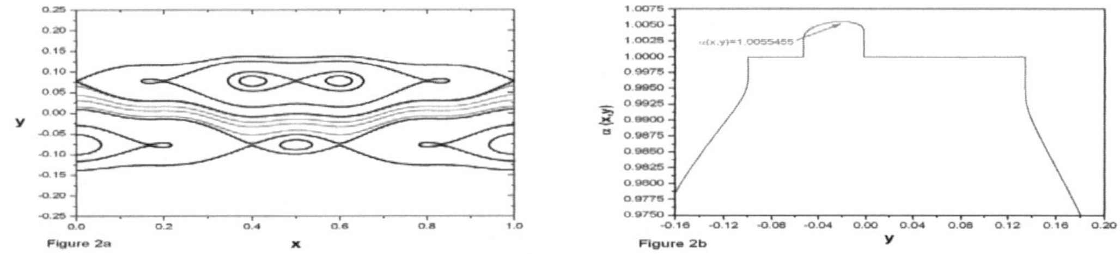

Figure 4: A meandering region (pink) is shown before the overlap of the isochronous resonances and the rotation numbers of the phase space in the left. The extreme of the rotation number corresponds to the shearless torus.

The parameter b controls the amplitude of the sinusoidal perturbations and allows the resonances overlapping. In figure (5) we chose b = 0.005. The coloured regions correspond to three different regions with meandering tori which are visualised through the three minima in the rotation number plot. The reader should note that for the current set of parameters the system seems to be integrable because the chaotic layers are so thin for the scale of the plots.

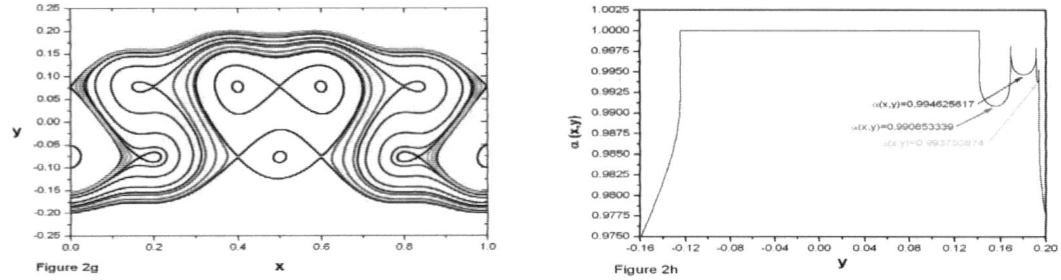

Figure 5: Three meandering regions are shown after the overlap of the isochronous resonances. There are three shearless curves each one associated with a minimum of the rotation number.

In figure (6) we increase the value of the perturbation parameter, now b = 0.043 and a wide chaotic region can be seen. The blue and green regions have been broken and the corresponding orbits became chaotic. The survivor red region contains the more robust meandering torus, called shearless curve, and when we look for its rotation number we observe that the red shearless torus has the lowest value among the three minima.

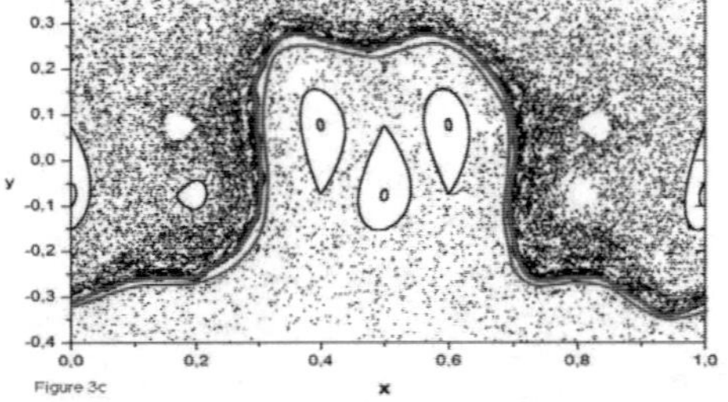

Figure 6: The destruction of the resonance structures and of the meandering regions is obtained increasing the parameter b. The more robust shearless, that has been survived to the effect of the perturbation, corresponds to the lowest minima of the rotation number.

5. Final remarks

We presented some fundamental tools of non-linear dynamics which are typical of Hamiltonian systems. We showed the isochronous resonances, which occur only in non-twist systems, governed by a Hamiltonian or by a map, robust tori which correspond to transport barriers for the dynamics and the meandering tori. These appear in the regions of isochronous resonances, as before as after the reconnection process. Even though the robust tori can occur as in twist as in non-twist systems, we presented their effects in a system with isochronous resonances. The movement of the robust tori among the island chains leads to two different topological rearrangements, one with the similar fixed points aligned and another one with the fixed points defocused. The meandering tori appear when the rotation number presents an extreme and we showed that the more robust of them has the minor minimum value. And to conclude we state that the isochronous resonances are actually part of the foundations of the non-linear theory.

Acknowledgements: We thank the Brazilian scientific agency, FAPESP, for financial support through the process 2010/ 20276-2.

References

[1] Chierchia L and Mather J N 2010 Scholarpedia **5**(9) 2123.
[2] Mackay R S and Meiss J D 1987 *Hamiltonian Dynamical Systems-a reprint selection*, Adam Hilger.
[3] Voyatziz G, Ichtiaroglou S 1999 Int. J. Bifurcation Chaos **9** 849.
[4] Egydio de Carvalho R and Ozorio de Almeida A M 1992 Phys. Lett. A **162** 457.
[5] Howard J E and Hohs S M 1984 Phys. Rev. E **29** 418.
[6] Delshams A and de la Llave R 2000 SIAM J. Math. Anal. **31** 1235.
[7] Ozorio de Almeida A M 1987 *Sistemas Hamiltonianos: Caos e Quantização* Editora da Unicamp.
[8] Egydio de Carvalho R 1993 Nonlinearity **6** 973.
[9] Voyatzis G, Meletlidou E and Ichtiaroglou S 2002 Chaos, Sol. Frac. **14** 1179.
[10] Egydio de Carvalho R, Martins, C GL and Favaro G M 2009 Braz. J. Phys. **39** 606.
[11] Egydio de Carvalho R and Favaro G M 2005 Physica A **350** 173.
[12] Martins C G L, Egydio de Carvalho R, Caldas I L, Roberto M 2011 Physica A **390** 957.
[13] del-Castillo-Negrete D, Greene J M and Morrison P J 1996 Physica D **91** 1.
[14] Wurm A, Apte A and Morrison P J 2004 Braz. J. Phys.**34** 1700.
[15] Szezech Jr J D, Caldas I L, Lopes S R, Viana R L and Morrison P J 2009 Chaos **19** 043108.
[16] del-Castillo-Negrete D and Morrison J P 1993 Phys. Fluids A **5** 948.
[17] Wurm A, Apte A, Fuchss K and Morrison J P 2005 Chaos **15** 023108.
[18] Martins C G L, Egydio de Carvalho R, Caldas I L and Roberto M 2011 J. Phys. A **44** 045102.

Studying the behaviour of averaged models in the third body perturbation problem

R C Domingos[1], A F Bertachini de Almeida Prado[1] and R Vilhena de Moraes[2]

[1]Instituto Nacional de Pesquisas Espaciais (INPE), São José dos Campos, SP, Brazil
[2]Universidade Federal de São Paulo (UNIFESP), São José dos Campos, SP, Brazil

E-mails: rdomingos95@gmail.com, prado@dem.inpe.br, rodolpho.vilhena@gmail.com

Abstract. The main goal of the present paper is to perform a study of the behaviour of the single and double averaged methods used to study the third-body perturbation problem in celestial mechanics. Both models are compared with the full elliptic restricted three-body problem. The study shows the differences between the values of the eccentricity and inclination of the satellite predicted by those models, showing the differences at every instant of time, as well as the integral over the time of the differences between the models. This integral is a new way to look at those differences and can give us a more complete view of the quality of the approximations, instead of just instantaneous views.

1. Introduction

There are many researches in the literature studying the third-body perturbation in a satellite in orbit around the Earth. Among the first ones are Kozai [1] that developed long-period and secular terms due to the perturbing potential of the Sun and the Moon. Cook [2] used the Lagrange's planetary equations of motion to study the long-period terms of the Sun and the Moon. Later, Kozai [3] considered the secular terms for asteroids with high inclination and eccentricity, considering the perturbation from Jupiter. Kaula [4] also studied this problem, obtaining new equations to calculate those effects. After that, Giacaglia [5] studied this problem with ecliptic elements for the Moon and equatorial elements for the satellite. Kozai [6] worked again in this problem and obtained the short period terms in closed forms and the secular and long period terms based in numerical integrations. Collins and Cefola [7] applied double averaged techniques to study satellites in high altitudes around the Earth, in particular for a long time span. Some other papers using the double average technique can be seen in Broucke [8], for expansions of second order, and Prado [9], for expansions up to the fourth order. Some other papers studied other planetary systems, like Carvalho ([10], [11]), Lara [12], Paskowitz and Scheers ([13], [14]). Domingos et al. [15] and Xiaodong et al. [16] included an inclination between the perturbing body and the orbital plane of the satellite. The goal of the single average is to eliminate only short periodic terms of the satellite to give an understanding of the long-term stability of the orbits, instead of considering also an average over the motion of the disturbing body, as done by the double-averaged model. The present paper makes a comparison of the three models, assuming that the perturbing body is in an elliptical orbit.

2. Dynamical model

It is assumed the existence of a main body, with mass m_0, that is fixed in the center of the reference system *x-y-z*. The perturbing body, having mass *m'*, is in an elliptic orbit with keplerian elements: semi-major axis a', eccentricity e', inclination i', argument of periapsis ω', longitude of the ascending node Ω', and mean motion n'. The spacecraft is in an orbit given by the orbital elements: semi-major axis *a*, eccentricity *e*, inclination *i*, argument of periapsis ω, longitude of the ascending node Ω, and mean motion *n*.

Using an expansion in Legendre's polynomials (considering $r' >> r$), the second order part of the disturbing potential, considering an average over the eccentric anomaly of the spacecraft, is [15]:

$$\langle R_2\rangle = \frac{\mu' a^2 n'^2}{2}\left(\frac{a'}{r'}\right)^3\left\{\left(1+\frac{3}{2}e^2\right)\left[\frac{3}{2}\left(\alpha^2+\beta^2\right)-1\right]+\frac{15}{4}\left(\alpha^2-\beta^2\right)e^2\right\} \tag{1}$$

where $r' = \dfrac{a'(1-e'^2)}{1+e'\cos f'}$, $\cos f' = \cos M' + e'(\cos 2M' - 1) + \dfrac{9}{8}e'^2\left(\cos 3M' - \cos M'\right) + \ldots$

and $\sin f' = \sin M' + e'\sin 2M' + e'^2\left(\dfrac{9}{8}\sin 3M' - \dfrac{7}{8}\sin M'\right) + \ldots$

For the situation of considering elliptic orbits for the perturbing body, the values of α and β are given by $\alpha = \cos\omega\cos D - \cos i\sin\omega\sin D$, $\beta = -\sin\omega\cos D - \cos i\cos\omega\sin D$ and $D = \Omega - f' - \omega'$ [15]. The mean anomaly of the perturbing body M' is given by $M' = M'_0 + n't$ and the Lagrange's planetary equations of motion are [17]:

$$\frac{de}{dt} = K\frac{15}{4}\mu' n'^2\frac{e\sqrt{1-e^2}}{n}\left[\sin 2\omega\left(\cos^2 D - \cos^2 i\sin^2 D\right) - \cos i\cos 2\omega\sin 2D\right] \tag{2}$$

$$\frac{di}{dt} = K\frac{1}{\sin i\left[\mu a\left(1-e^2\right)\right]^{\frac{1}{2}}}\frac{3}{4}\mu' n'^2 a^2\left\{e^2\left[-5\cos 2\omega\sin 2D\cos^2 i + \frac{3}{2}\left(-\sin 2D + \sin 2D\cos^2 i\right) + \right.\right.$$
$$\left.\left. -\frac{5}{2}\sin D\cos 2\omega(1+\cos^2 i)\right] + \left(\sin D - \sin 2D\cos^2 i\right) - 5\cos 2D\sin 2\omega\cos i\right\} \tag{3}$$

$$\frac{d\Omega}{dt} = K\frac{3}{4}\mu' n'^2\frac{a^2}{\sin i\left[\mu a\left(1-e^2\right)\right]^{\frac{1}{2}}}\left\{\left(1+\frac{3}{2}e^2\right)\left(-\sin^2 D\sin 2i\right) + \left[\cos 2i\cos 2\omega\sin^2 D + \right.\right.$$
$$\left.\left. + \sin i\sin 2D\sin 2\omega\right]\left(\frac{5}{2}e^2\right)\right\} \tag{4}$$

$$\frac{d\omega}{dt} = K\frac{3}{4}\mu' n'^2\left\{-\frac{a^2}{\sin i\left[\mu a\left(1-e^2\right)\right]^{1/2}}\left[\left(1+\frac{3}{2}e^2\right)\left(-\sin^2 D\sin 2i\right) + \right.\right.$$
$$\left. + \left(\frac{5}{2}e^2\right)\left(\cos 2i\sin^2 D\cos 2\omega\right) + \frac{1}{2}\sin 2i\sin 2D\sin 2\omega\right] +$$
$$\left. + \frac{\sqrt{1-e^2}}{n}\left[5\left(\cos^2 D - \cos^2 i\sin^2 D\right)\cos 2\omega - 5\cos i\sin 2D\sin 2\omega + \frac{6}{2}\left(\cos^2 D + \cos^2 i\sin^2 D\right)\right]\right\} \tag{5}$$

Journal of Physics: Conference Series **465** (2013) 012017 doi:10.1088/1742-6596/465/1/012017

where $K = \dfrac{1+3e'\cos f' + 3e'^2\cos^2 f'}{1-6e'+15e'^2}$

3. Numerical results

The high-altitude circular orbits, assumed to be an orbit with semi-major axis of 42000 km in the present paper, near the geostationary orbits, are the ones more affected by the third body perturbations, so a detailed study is made regarding the comparisons among the different models. This task is done by choosing a specific value for the initial inclination and then analyzing the evolution of the magnitude of the differences between the values obtained by the two averaged models mentioned here and the full restricted elliptic three-body problem. The instantaneous errors between each of the averaged models and the full model are plotted as a function of the eccentricity and inclination of the orbit of the spacecraft. Then, the integral of the magnitude of these differences over the time is also obtained. This quantity shows better how far the curves are from each other taking into account the whole interval of time and not focusing only in individual points. All the comparisons are made for a short time period of two days (which correspond to ~0.46 canonical units), because the idea is to see the difference of the models for shorter times.

Figure 1 shows the results for an initial inclination of 20 degrees, that is a value below the critical angle. The dashed line represents the results obtained by the double average model and the continuous line represents the single averaged model. Eccentricities of the primaries of 0.1 are shown in red, 0.2 in green and 0.3 in blue. The circular case is not shown here because the results are too close among the models and eccentricities higher than 0.3 are also omitted because the approximations are not good. Note the very small errors, in particular for the inclination, for both averaged models.

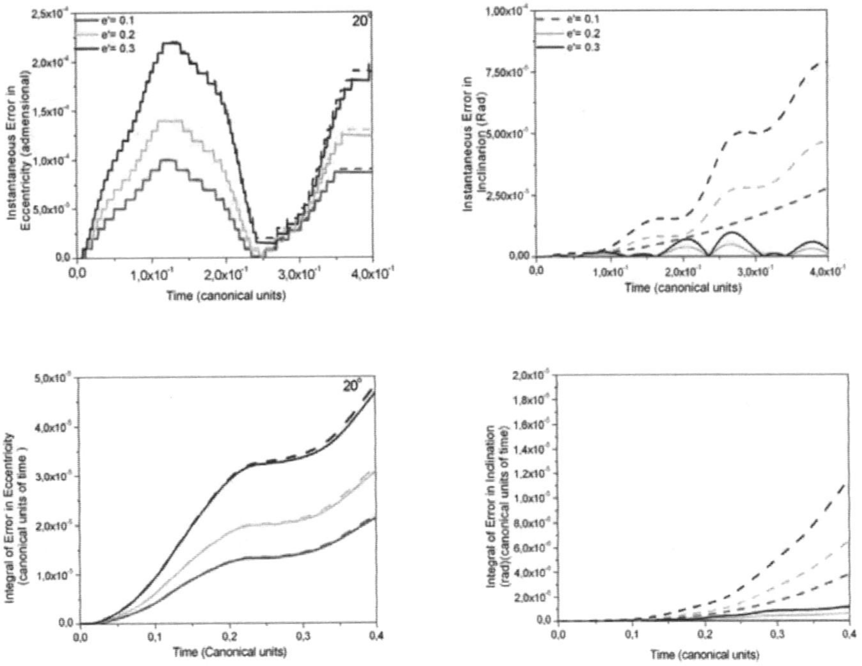

Figure 1. Instantaneous and integral errors in eccentricity and inclination as a function of time for High-Earth orbits with initial inclination of 20°. The eccentricity of the perturbing body (e') is shown in colours codes: 0.1 (red), 0.2 (green) and 0.3 (blue). The lines are: single-averaged model (solid line) and double-averaged model (dashed line).

It is visible that the quality of the results decreases when the orbit of the primaries become more eccentric. The magnitude of the error has shorter period for the inclination, when compared to the eccentricity. It is visible that the single-averaged model is better than the double-averaged model for shorter times, as expected, since it takes into account the short period motion of the spacecraft.

Figure 2 shows the results for an initial inclination of 40 degrees, which is very near the critical angle. This type of analyses confirms the ones made before. Note the decrease of the quality of the second order averaged approximations with the increase of the eccentricity of the primaries. It is also noted and quantified how this quality is reduced with time. Note also the oscillations of the errors, also with shorter period for the error in inclination. It means that the second order averaged approximations loose accuracy, recovered it again and so on. The averaged models are very good when e' = 0.1. The single-averaged model is again better than the double-averaged model, for the same reasons explained before.

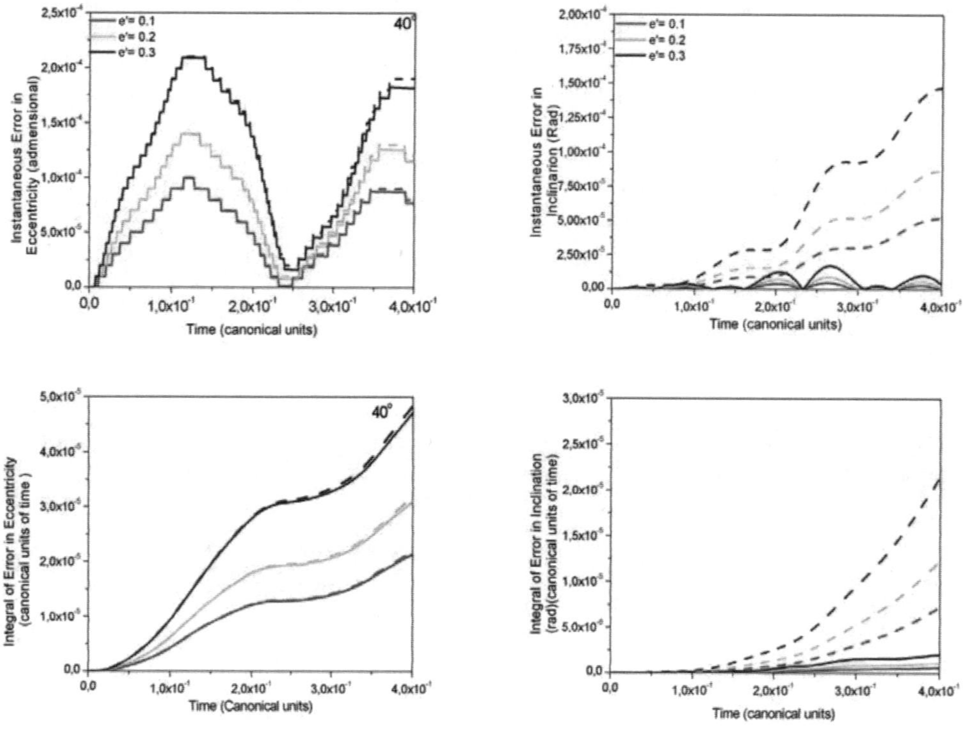

Figure 2. Instantaneous and integral errors in eccentricity and inclination as a function of time for High-Earth orbits and initial inclinations of 40°. The eccentricity of the perturbing body (e') is shown in colour codes: 0.1 (red), 0.2 (green) and 0.3 (blue). The lines are: single averaged model (solid line) and double-averaged model (dashed line).

Figure 3 shows the results for an initial inclination of 60 degrees, which is above the critical angle. Of course the decrease of the quality of the second order averaged approximations with the increase of the eccentricity of the primaries is also evident by this type of analysis. The oscillations of the errors also occur for this inclination. Again, the averaged models are excellent when e' = 0.1 and the double averaged model is better than the single-averaged model.

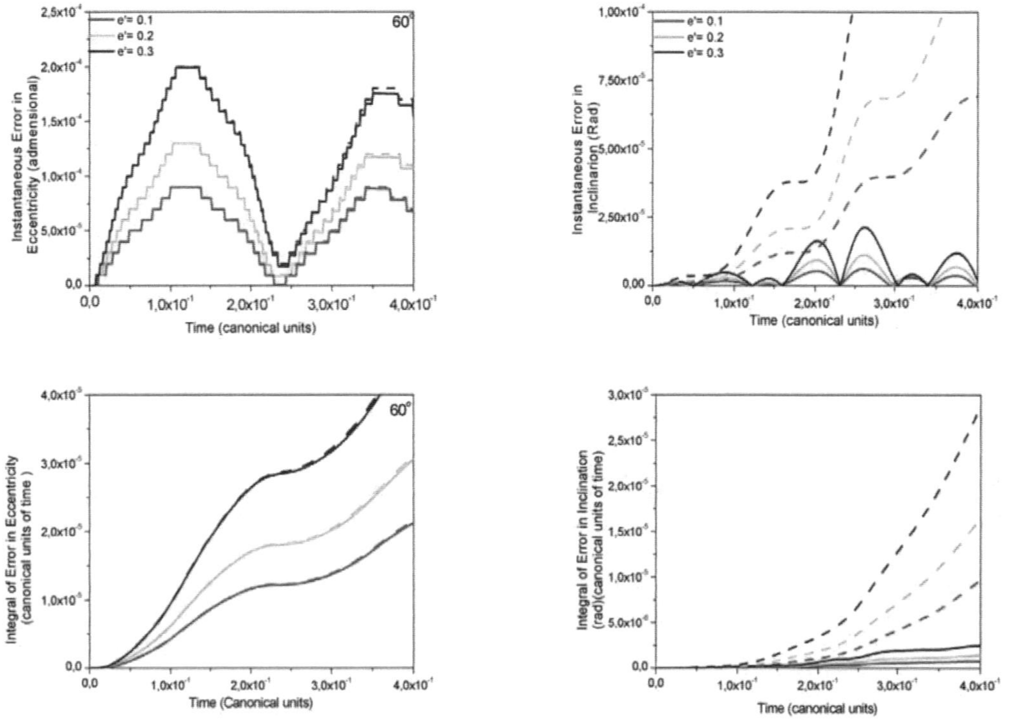

Figure 3. Instantaneous and integral errors in eccentricity and inclination as a function of time for *High-Earth Orbits* and initial inclinations of 60°. The eccentricity of the perturbing body (e') is shown in color codes: 0.1 (red), 0.2 (green) and 0.3 (blue). The lines are: single average model (solid line) and double averaged model (dashed line).

4. Conclusions

A comparison of the double and single-averaged models up to the second-order with the full restricted elliptic problem was performed. The second order averaged models are accurate for a perturbing body in a circular orbit. This accuracy decreases with the increase of the eccentricity of the perturbing body. The second order averaged models are not accurate and should not be used for eccentricities of the perturbing body of 0.3 or larger.

Then, a new approach that consists of studying the integral of the errors between the models over the time was made. It showed and quantified the better accuracy of the single-averaged model for the time span of two days used in the present research. This approach showed to be an interesting form of quantifying those differences.

Acknowledgment

The authors wish to express their appreciation for the support provided by CNPq under the contracts N° 150195/2012-5 and 304700/2009-6, FAPESP (contracts 2011/09310-7, 2011/08171-3 and 2012/21023-6) and CAPES.

References

[1] Kozai Y. On the effects of the Sun and Moon upon the motion of a close Earth satellite, Smithsonian Inst. Astrophysical Observatory (1959), Special Report. 22.
[2] G. E. Cook. Luni-solar perturbations of the orbit of an earth satellite, The Geophysical Journal

Journal of Physics: Conference Series **465** (2013) 012017 doi:10.1088/1742-6596/465/1/012017

of the Royal Astronomical Society, 6 (1962), n. 3, 271.
[3] Kozai Y 1962 *Astron. J.* **67** 591
[4] W.M. Kaula W M 1962 *Astron. J.* **67** 300
[5] G.E.O. Giacaglia. Lunar perturbations on artificial satellites of the earth, Smithsonian Astrophysical Observatory (1973), Special Report 352, 1.
[6] Y. Kozai. A new method to compute lunisolar perturbations in satellite motions, Smithsonian Astrophysical Observatory (1973), Special Report 349, 27.
[7] S.K Collins, P.J Cefola. Double averaged third body model for prediction of super-synchronous orbits over long time spans, AIAA, 64, (1979).
[8] Broucke R A 2003 *J. Guid. Control Dyn.*, **26 (1)** 27
[9] Prado A F B A 2003 *J. Guid. Control Dyn.* **26 (1)** 33
[10] Carvalho J P S, ELIPE A, Vilhena de Moraes R and A.F.B.A. Prado A F B A 2010. *Celest. Mech. Dyn. Astron* **108** 371.
[11] J.P.S. Carvalho J P S, ELIPE A, Vilhena de Moraes R and Prado A F B A 2011 *Math. Probl. Eng.* **2011** 1.
[12] M. Lara. Design of long-lifetime lunar orbits: A hybrid approach. Acta Astronaut. 69(3–4), 186–199 (2011).
[13] M.E Paskowitz M E and Scheeres D J 2006 *J. Guid. Control Dyn.* **29(5)** 1147
[14] Paskowitz M E and Scheeres D J 2006 *J. Guid. Control Dyn.* **29(2)** 342
[15] Domingos R C, Vilhena de Moraes R and Prado A F B A. 2008 *Math. Probl. Eng.* **2008a** 1
[16] Xiaodong L, B. Hexi B and Xingrui M 2012 *Astrophys. Space. Sci.* **339** 295
[17] Domingos R C, Prado A F B A and Vilhena de Moraes R 2013 *Math. Probl. Eng.***2013**.

Journal of Physics: Conference Series **465** (2013) 012018 doi:10.1088/1742-6596/465/1/012018

Exploring sensitive dependence and transitivity to optimize travel time in chaotic systems

S H S Santana[1], E E N Macau[1], O C Winter[2] and L F A França[1]
[1]Instituto Nacional de Pesquisas Espaciais - INPE, Avenida dos Astronautas, 1758 - CEP 12227-010, São José dos Campos, SP, Brazil
[2]Universidade Estadual Paulista - UNESP, Grupo de Dinâmica Orbital e Planetologia, Avenida Ariberto Pereira da Cunha, 333 - CEP 12516-410, Guaratinguetá, SP, Brazil

E-mail: saymonhss@gmail.com

Abstract. Transitivity and sensitive dependence on initial conditions are the main characteristics of chaotic behavior. The latter one can be exploited so that small controlled perturbations in system parameters may imply a faster transfer in time from a desired start point to a neighborhood of a desired final state. In this study three targeting approaches are evaluated: The first one uses a geometric approach to find the proper perturbation which allows a faster transfer between two desired points; The second, an evolutionary algorithm called GEO (Generalized External Optimization), is adapted to search for optimized orbits; The third one, uses successive perturbations along the path in order to direct the orbits to the final desired point in a short time interval. These three methods are evaluated regarding performance and implementation complexity.

1. Introduction

Since the 18th century science has expended efforts to understand the behavior of nonlinear systems. Several systems in nature, such as the n-body problem in celestial mechanics, exhibit chaotic behavior.

Although notable scientists have worked to understand the behavior of systems of celestial mechanics, only in 1890, Poincaré showed that in the presence of homoclinic intersections, there is a sensitive dependence on initial conditions.

Chaotic systems are sensitive to initial conditions. This property, may be used in order to optimize the transport between two desired points of the system. Once the future state of a chaotic system may be substantially changed by small perturbations, as will be shown in our results, is possible to reduce the travel time of a path if a suitable perturbation is applied. This procedure, often called targeting, consists of finding a nearby short path using only avaliable dynamics. In references [6] [4] [5] one may see the applicability of this method in orbital dynamics.

Our study focuses on trajectories in a Hénon Map [3], which takes a point (x_n, y_n) in the plane and maps it to a new point (x_{n+1}, y_{n+1}).

$$x_{n+1} = 1 - \alpha x_n^2 + y_n \qquad (1)$$

$$y_{n+1} = \beta x_n \qquad (2)$$

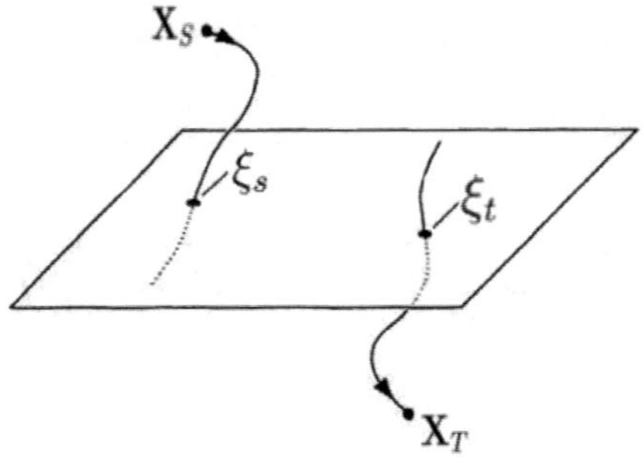

Figure 1. Surface of section from a three dimensional flow

For $\alpha = 1.4$ and $\beta = 0.3$ the map presents a chaotic behavior.

In this work, we analize trajectories starting from an initial state (x_I, y_I) (called source-point) to a final state (x_F, y_F) (target point). In order to reduce the time spent in that path, three targeting approachs were implemented. The first one [7] uses a geometric approach to find the proper perturbation which allows a quick transfer between two desired points; The second, [2] an evolutionary algorithm called GEO (Generalized External Optimization), is adapted to search for optimized orbits; The third one [1] uses successive perturbations along the path in order to direct the orbits to their final desired point in a short time interval. Each one of these methods are explained in the following sections.

2. Geometric Approach

Consider a three-dimensional continuous-time dynamical system $\frac{d\mathbf{X}}{dt} = \mathbf{F}(\mathbf{X})$ and employing a surface of section whose coordinates are denoted by ξ and the Poincaré map $\xi_{n+1} = \mathbf{f}(\xi_n, \alpha)$, where $\mathbf{f}$ is an invertible map and α a system parameter.

Suppose one wish to go from a source point $\mathbf{X}_S$ to a small region around a target point $\mathbf{X}_T$ (as shown in figure 1).

We assume that the system's parameter α is available for adjustment at each iterate, such that $\alpha_n = \overline{\alpha} + \delta_n$, where $\overline{\alpha}$ is a nominal value of α and δ_n denotes a small deviation from $\overline{\alpha}$.

After one iteration of the return map, the change of state δ_ξ due to a small perturbation δ_1 relative to the point $f(\xi_s, \overline{\alpha})$ is given by:

$$\delta\xi \cong \delta_1 \frac{\partial f(\xi_s, \alpha)}{\partial \alpha}\bigg|_{(\overline{\alpha})} \tag{3}$$

where δ_1 varies through a small interval $|\delta_1| < \delta_*$.

The method consists in iterating the segment $\overline{\delta\xi}$ and the region $\overline{\epsilon_T}$ around the target, forward and backward respectively, using $\alpha_n = \overline{\alpha}$, until it is detected the intersection of the images of $\overline{\delta\xi}$ and $\overline{\epsilon_T}$. Thus, a point in $\overline{\delta\xi}$, near to the target point, which is mapped to the target region $\overline{\epsilon_T}$ in $n_1 + n_2$ iterations, is found, where n_1 (n_2) is the number of iterations of the segment $\delta\xi$ (region $\overline{\epsilon_T}$).

In pratice, it is not possible to iterate either the segment $\overline{\delta\xi}$ or the region $\overline{\epsilon_T}$. Actually, we iterate discrete approximations to them (See figure 2) and make successive refinements, joining their images with straight-line segments, until a sufficiently accurate solution has been obtained.

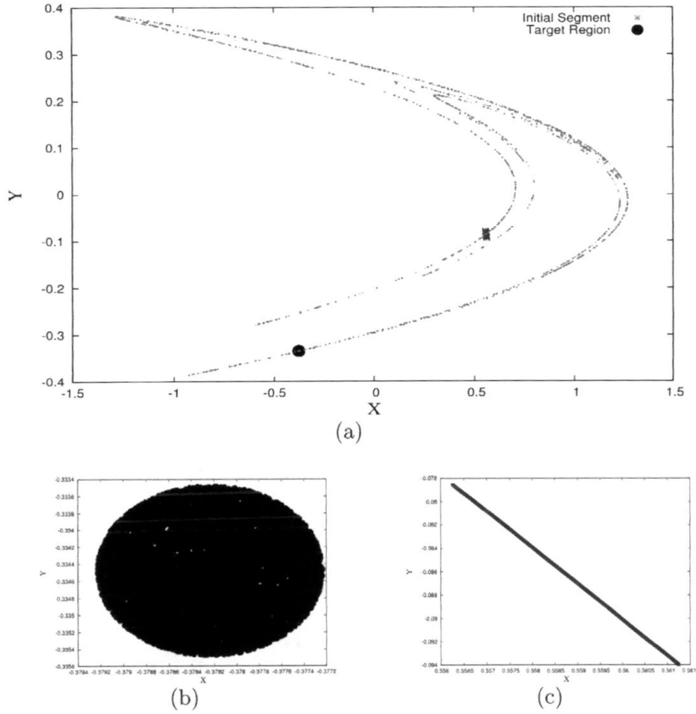

Figure 2. A trajectory in Hénon Map for $\alpha = 1.4$ and $\beta = 0.3$ is shown in (a), the blue points represent the initial segment $\overline{\delta\xi}$ and the black ones the target region $\overline{\epsilon_T}$. In (b) and (c) zoomed images of $\overline{\delta\xi}$ and $\overline{\epsilon_T}$ are shown, respectively.

3. Generalized Extremal Optimization

GEO is an evolutionary algorithm in which, unlike other bio-inspired algorithms, there is not a population of strings (or solutions), but a population of bits represented by a single string as shown below. Moreover, there is no crossing between individuals. Instead, each bit in GEO (species) is "forced" to mutate with a probability proportional to its assigned fitness.

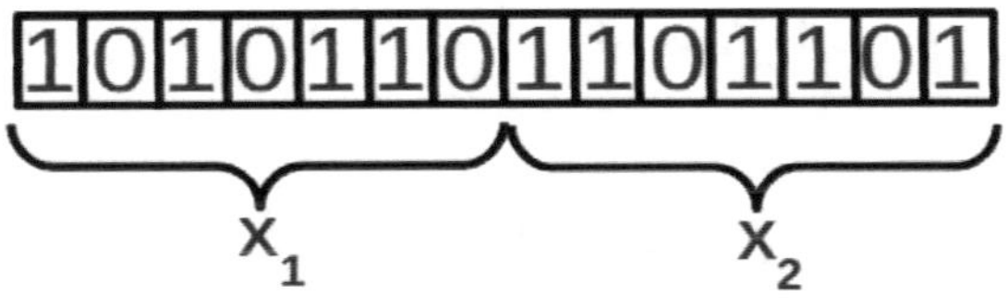

Figure 3. Design variables X_1 and X_2 encoded in a binary string

In this work GEO was used in order to find a point (called optimal point) near to the source, from which the number of iterations needed to reach a region in the target's neighborhood is minimized.

The practical implementation of GEO algorithm is written as

(i) Initialize randomly a binary string of length L which encodes N design variables of bit length $l_j (j = 1, ..., N)$. For the initial configuration C of bits, calculate the objective function value V and set $C_{best} = C$ and $V_{best} = V$.

(ii) For each bit i of the string, at a given iteration:
   (a) Flip the bit and calculate the objective function value V_i of the string configuration C_i,
   (b) Set the bit fitness as $\Delta V_i = (V_i - V_{best})$. It indicates the relative gain (or loss) which one has when the bit is mutated, compared to the best objective function value found so far.
   (c) Return the bit to its original value.

(iii) Rank the bits according to their fitness values, from $k = 1$ for the least adapted bit to $k = L$ for the best adapted. In a minimization problem, higher values of V_i will have higher ranking, whereas for maximization problems is the opposite. If two or more bits have the same fitness, rank them randomly with uniform distribution.

(iv) Choose with equal probability a candidate bit i to mutate. Generate a random number RAN with uniform distribution in the range [0,1]. If the probability of mutation $P_i(k) = k^{-\tau}$ (where τ is an adjustable positive parameter) of the chosen bit is equal or greater than RAN the bit is confirmed to mutate. Otherwise, the process is repeated until a bit is confirmed to mutate.

(v) For the mutated bit set $C = C_i$ and $V = V_i$.

(vi) If $V < V_{best}$ ($V > V_{best}$, for a maximization problem) then set $V_{best} = V$ and $C_{best} = C$.

(vii) Repeat steps 2 to 6 until a given stopping criterion is fulfilled.

(viii) Return C_{best} and V_{best}.

4. Successive Perturbations

Consider a map, given by $Z_{n+1} = F(b, Z_n)$, where b is a parameter which can be changed by $\pm\delta$, so that it can assume three possible values: $b + \delta$, b e $b - \delta$. The method uses the three possible parameter values to direct a trajectory from a starting point P_i to a neighborhood of target P_f.

Initially, the map F is applied to the point P_i, using three possible values of b. The first iteration generates, then, three new points: $Z_{1,1} = F(b + \delta, Z_0)$, $Z_{1,2} = F(b, Z_0)$ and $Z_{1,3} = F(b - \delta, Z_0)$.

This procedure is kept with $Z_{1,1}$, $Z_{1,2}$ e $Z_{1,3}$. Thus, after the second iteration, using the possible values of b, nine points are generated. The process is repeated until one of points $Z_{i,j}$ reaches the neighborhood of the final desired neighborhood.

In the following results the successive perturbations are represented, respectively, by a sequence specifying which disturbance was applied at each iteration, such that: $-\delta$ indicates $\alpha = \bar{\alpha} + \delta$, 0 indicates $\alpha = \bar{\alpha}$, $+\delta$ indicates $\alpha = \bar{\alpha} - \delta$

The possibles paths from a point Z_0 are shown in figure 4.

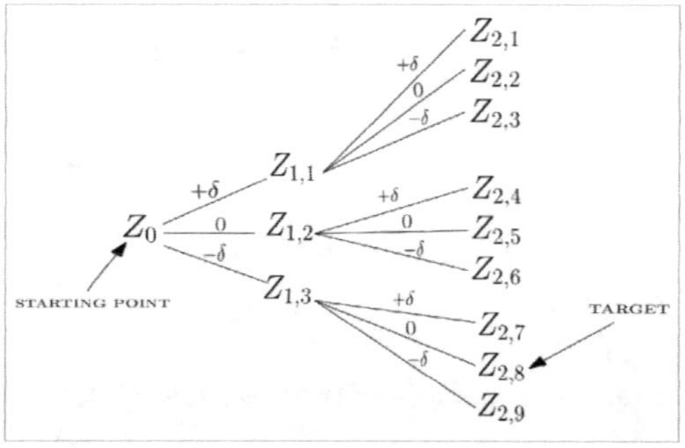

Figure 4. Possible paths to reach the target

5. Results

In all trajectory shown below, the original orbit reaches the target's neighborhood after 10^6 iterations.

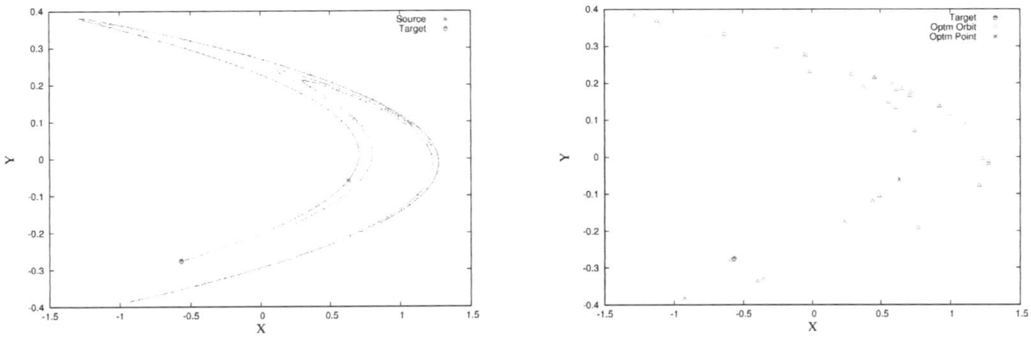

Figure 5. Geometric approach. Optimal orbit obtained with 37 iterations. Required perturbation $\delta(x, y) = (-3.109 \times 10^{-2}, 5.195 \times 10^{-4})$

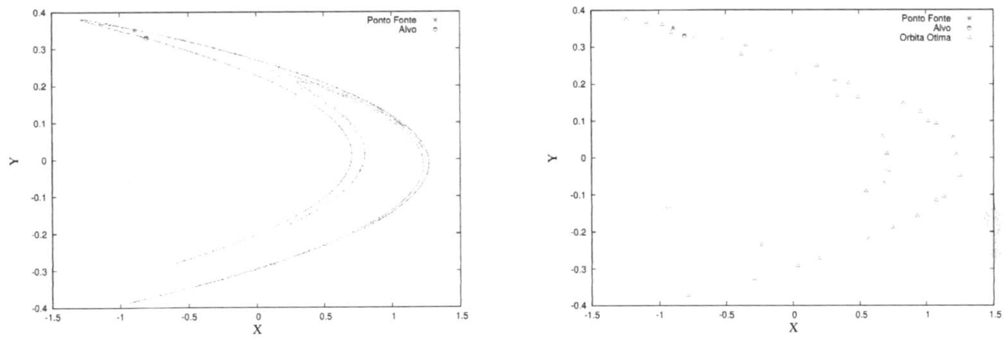

Figure 6. Generalized Extremal Optimization. Optimal orbit obtained with 36 iterations. Required perturbation $\delta(x, y) = (-1.416 \times 10^{-2}, -6.069 \times 10^{-2})$

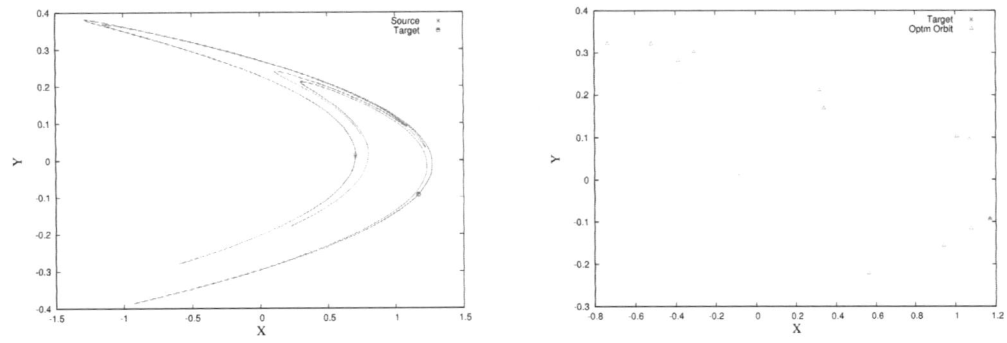

Figure 7. Successive Perturbations. Optimal orbit obtained with 12 iterations. Sequence of perturbations: $-\delta, 0, +\delta, 0, -\delta, -\delta, 0, +\delta, -\delta, -\delta, -\delta, 0$

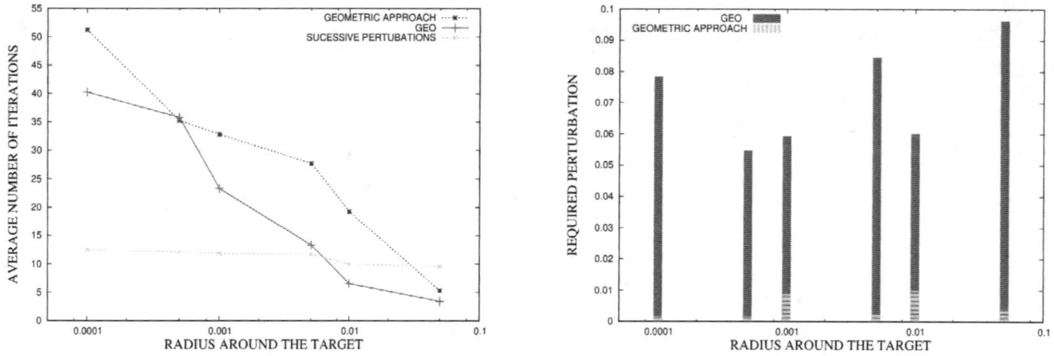

Figure 8. Dependence of the average number of iterations with the target's Radius (a). Comparison between the required pertubation *versus* target's radius for GEO and Geometric Approach methods (b). For each method, 25 trajectories were simulated for different values of radius of the target's neighborhood.

6. Final Remarks

As GEO ignores the dynamics of the system, unknowing the feasibles regions, although the number of iterations is (on average) less than the found for the geometric approach, in some cases the optimal point may be located away from basin of attraction, thus requiring a larger perturbation.

Geometric approach, on the other hand, found, on average, the optimal trajectory within a higher number of iterations. It was able, however, to find points quite close to the source point, requiring a perturbation in order one or two times lower than the one found in GEO or successive perturbations.

The method of successive perturbations may be adjusted so that tiny values of δ are used, however, at each level N, 3^N analyses are performed, hence the required memory may be computationally impracticable.

In the presented results it is possible to note the feasibility of using sensitive dependence to guide a trajectory to a final desired state quickly. The implemented methods may also be applied to higher dimention systems.

References

[1] Baptista, M. S.; Caldas, I. L. - **Easy-to-Implemente Method to Target Nonlinear Systems** - *CHAOS.* 1998. v 8, p 290.

[2] De Sousa, F. L.; Ramos, F. M. - **New Stochastic Algorithm for Design Optimization** - *AIAA Journal.* 2003. v 41, p 1808.

[3] Hénon, M. - **A Two-Dimensional Mapping with a Strange Attractor** - *Communications in Mathematical-Physics.* 1976. v 50, p 69.

[4] Macau, E. E. N.; Caldas, I. L. - **Driving trajectories in chaotic scattering** - *Physical Review E.* 2002. v 65, p 2621.

[5] Macau, E. E. N.; De Sousa, F. L.; Ramos, F. M. - **Using Chaos to Guide a Spacecraft to the moon** - *Acta Astronautica.* 2005. v 47, p 871.

[6] Macau, E. E. N.; Grebogi C. - **Control of chaos and its relevancy to spacecraft steering** - *Philosophical Transactions. Royal Society. Mathematical, Physical and Engineering Sciences.* 2006. v 364, p 2463.

[7] Shinbrot, T.; Ott, E.; Grebogi, C.; Yorke, J. A. - **Using Chaos to Direct Trajectories to Targets** - *Physical Review Letters.* 1990. v 65, p 3215.

A Study of the Duration of the Passage through the Van Allen Belts for a Spacecraft going to the Moon

T C Oliveira[1], E M Rocco[2], A F B A Prado[3], J L Ferreira[4]
[123]Space Mechanics and Control Division, National Institute for Space Research - INPE, São José dos Campos-SP, BR.
[4]Universidade de Brasília, Campus Darcy Ribeiro, Plano Piloto, 70910-900 Asa Norte, DF, Brazil

E-mail: [1]thais.tata@gmail.com, [2]evandro@dem.inpe.br; [3]prado@dem.inpe.br; [4]leo@fis.unb.br

Abstract. This paper has the goal of estimating the fuel consumption and the duration of the transit in the Van Allen belts for a flight of a spacecraft going from the Earth to the Moon. This problem is very important because the region interior to the belts have a high density of energetic charged particles that can damage the satellite, so minimizing this transit time helps in protecting the equipments on board. The propulsive force is assumed to have a low magnitude and to be applied in the direction of the motion of the spacecraft to maximize the energy transferred to the space vehicle. Perturbation forces are considered in the dynamical model and they influence in both results, consumption and transit time.

1. Introduction

This paper studies maneuvers for a spacecraft going from the Earth to the Moon, in particular making estimations of the time of flight that the spacecraft stays inside the Van Allen belts and the fuel consumption to perform the maneuvers. This research is a first approach to study this problem, which depends on the propulsion system available.

The existence of the Van Allen belts was first proposed by James Van Allen, in 1958, and after that satellites launched in the same year confirmed their existence [1]. The Van Allen belts can be described as magnetic mirrors. The magnetic field of the Earth, with strong intensity on the poles and weak intensity in the equator, forms a natural magnetic mirror that confines high energetic electrons and ions [2].

There is a great importance in studying these belts, since the confined energetic particles can damage the components of the satellites and also humans travelling in space. In this way, it is important for missions that pass through these belts the minimization of the radiation exposure and the study of how much radiation the satellite can receive. In this way, the search for trajectories that minimizes the time that the satellite stays inside these belts is an important part of the mission design.

The Van Allen belts are not yet fully understood and there are still today new discoveries related to them. Two identical spacecrafts were launched in August 2012, named the Van Allen Probes, with the objective of studying the Van Allen belts. These probes have made some important discoveries about the Van Allen Belts and scientists hope they will be able to make new discoveries to understand it better in the near future [3].

2. The Spacecraft Trajectory Simulator

In order to study these maneuvers, the Spacecraft Trajectory Simulator was used. This simulator is a complex package that was developed to help mission designers to plan space missions. It is capable of simulating orbital maneuvers in a realistic environment, considering the perturbations of the Sun, the Moon, the geopotential and also the radiation pressure. It is also possible to consider errors, delays and failures in the sensors and actuators. This simulator considers a satellite in a closed-loop system with Proportional-Integral-Derivative (PID) controller. The greatest advantage of using this package in the present study is the possibility of simulating the maneuvers of the spacecraft going to the Moon in this realistic environment [4]. The spacecraft starts its motion in a circular orbit around the Earth and the goal is to send it to an orbit around the Moon. In the phase of the mission studied here, the final constraint is not that the satellite enters in orbit around the Moon or that it passes by the Moon. The constraint is just that the orbit of the spacecraft has an apogee at the altitude of the orbit of the Moon, to make an encounter possible. The perturbations considered were the J_2, J_3 and J_4 terms of the geopotential, the third-body perturbation due to the Sun and the Moon and the solar radiation pressure. The spacecraft used in the simulations has a mass of 150 kg and the area of the object exposed to the solar radiation is 1 m^2. The initial orbit has the following orbital parameters: semi-major axis = 7178 km; eccentricity = 0; inclination = 0, 45 and 90 degrees, right ascension of the ascending node = 0 degrees; mean anomaly = 0 degrees. The initial date and time of the simulation is 01/01/2010 at 05h30m00s. The reason to study different values for the inclination of the orbit of the spacecraft is that this parameter has a large impact in the transit time by the Vah Allen belts, since the belts are confined in the latitudes of 65 degrees north and south, so inclinations larger than 65 degrees will have the advantage of having part of its trajectory outside the belts.

3. The Propulsion System

The propulsion systems considered in this paper were two different variations of the hall thrusters that are being under development at the UnB (Universidade de Brasilia) [5]. A Hall thruster is a kind of ion thruster that uses the electric field to accelerate the propellant. The propellant itself is composed by ions that are accelerated to high speeds. More information about the working principles of the hall thrusters can be found in reference [5]. The two types of hall thrusters used in the simulations were the PHALL I and PHALL II. These thrusters have different structures and parameters. Although this propulsion system was nor used yet for real missions, many papers studying the efficiency and optimal maneuvers was already developed, like shown in reference [6], where it is studied optimal maneuvers with minimum fuel consumption with the use of these PHALLs in order to achieve a final given orbit. The parameters of both propulsion systems are given bellow [5], in Table 1.

Table 1 – Comparative parameters of PHALL I and PHALL II.

	PHALL I	PHALL II
Average Thrust (mN)	120,0	84.9
Specific Impulse (s)	1600	1083

4. The Van Allen belts

The model of the Van Allen belts used in the present research is based in reference [7]. They are composed by two belts, the inner and the outer one. Both of them extend to 65 degrees of latitude in both north and south directions. The inner belt goes from 1000 km to 6000 km of altitude, while the outer belt extends from 13000 km to 60000 km of altitude. Figure 1 shows a sketch with their main characteristics.

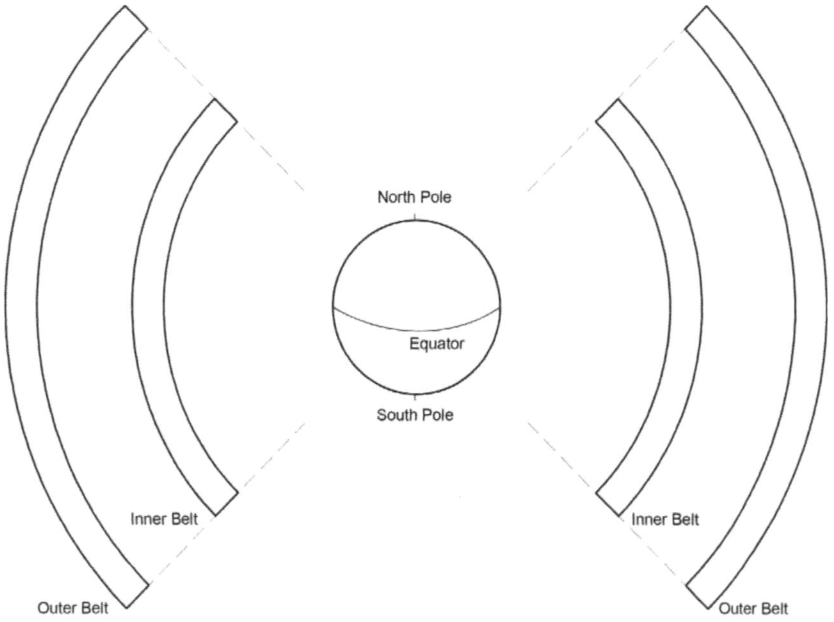

Fig. 1 - Geometry of the Van Allen belts.

5. Results

The results presented here are based on a maneuver that sends the spacecraft to an orbit with apogee of 384.400 km. The parameters of the initial orbit are given in section 2, the propulsion system is shown in section 3 and the direction of the propulsion applied to the spacecraft is tangential. The fuel consumption and the time spend in the Van Allen belt is shown in Table 1.

Table 1 – Results for different maneuvers

Propulsion System	Thrust capacity (N)	Inclination of the orbit (degrees)	Time spend in the Van Allen belts (days)	Fuel Comsumption (Kg)
Phall I (8 thrusters)	1 N	0	5.50	46.39
Phall I (8 thrusters)	1 N	45	5.63	47.33
Phall I (8 thrusters)	1 N	90	3.96	47.21
Phall I (4 thrusters)	0.5 N	0	10.96	47.23
Phall I (4 thrusters)	0.5 N	45	11.04	48.20
Phall I (4 thrusters)	0.5 N	90	8.09	48.06
Phall II (8 thrusters)	0.68 N	0	7.59	63.52
Phall II (8 thrusters)	0.68 N	45	7.61	64.7
Phall II (8 thrusters)	0.68 N	90	4.42	64.52
Phall II (4 thrusters)	0.34 N	0	15.15	64.47
Phall II (4 thrusters)	0.34 N	45	15.18	65.65
Phall II (4 thrusters)	0.34 N	90	10.90	65.50

From Table 1 it is possible to see several aspects of each maneuver. The maneuvers with inclination equal to 45 degrees were the worst maneuvers to consider for the mission. They spend more time in the Van Allen belts, due to the geometry of this orbit. The fuel consumption is also larger, if compared to the other maneuvers. The reason for this larger fuel consumption is the effects of the perturbation due to J_2, which is considered during the propelled phase of the maneuver. This perturbation dominates the system for a long time of the maneuver and it is stronger for orbits with 45 degrees of inclination, been smaller for equatorial orbits. Maneuvers with zero degrees of inclination (equatorial orbits) were the second ones in terms of the time spend on the Van Allen belts, but the fuel consumption were smaller than the others. For maneuvers with 90 degrees of inclination, the times spend on the Van Allen belts were minimum. This occurs because of the symmetry of the belts, which extends to 65 degrees north and south in latitude. The amount of fuel consumption for these maneuvers is between the values of the others.

Phall I is more efficient than Phal II, because of the higher specific impulse and also the higher magnitude of the thrust. The times spend in the Van Allen belts when using Phall I were smaller with respect to the same maneuvers made using Phall II, and also the fuel consumption were smaller. The largest time obtained in the Van Allen belts were slightly more than 15 days, and it occurred for the maneuver realized with 4 Phall II thrusters and considering an inclination equal to 45 degrees.

Next, it is shown some figures for the maneuver realized with 8 Phall I thrusters, shown in the first line of Table I. The inclination of this maneuver is zero, so a planar view of the trajectory is shown here, in an inertial reference frame with the Earth in the center and the reference plane passing by the equator.

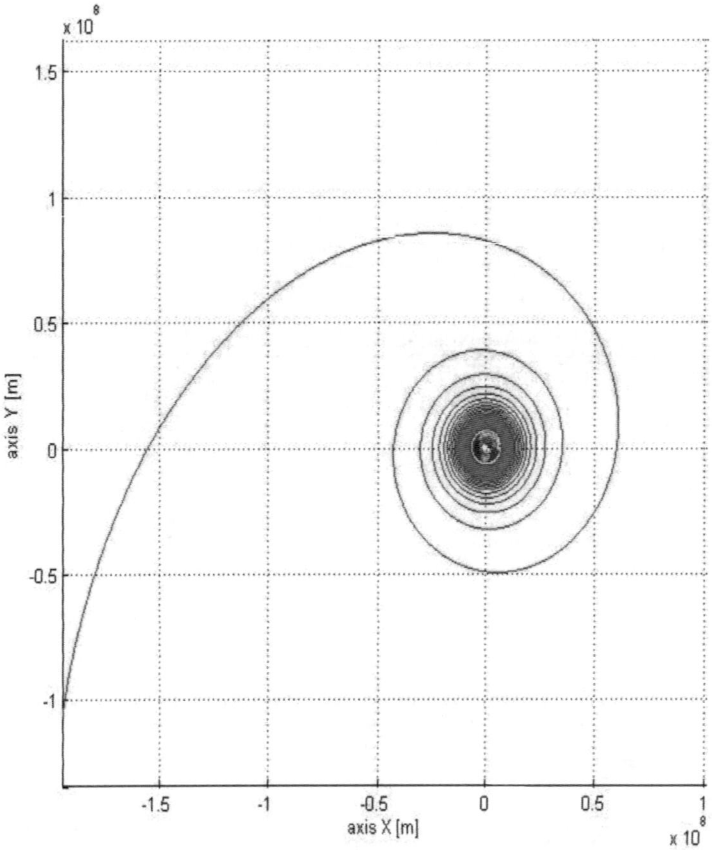

Fig. 2 – Maneuver realized with 8 Phall I thrusters as the propulsion system.

From Figure 2 it is possible to see the orbit of the satellite gaining velocity until it achieves the final orbital condition. Figure 3 shows the thrust applied in each axis of the inertial frame and the total thrust applied, which in this case is one Newton. This figure shows how the control system works to perform the maneuver, by specifying each component of the thrust force applied to the spacecraft, so given the solution of the problem in terms of the force vector. There is no component of the force in the z-axis, since the maneuver is equatorial. Figure 4 shows the fuel consumption as a function of the time. It is possible to note that the fuel consumption increases linearly as the maneuver goes by.

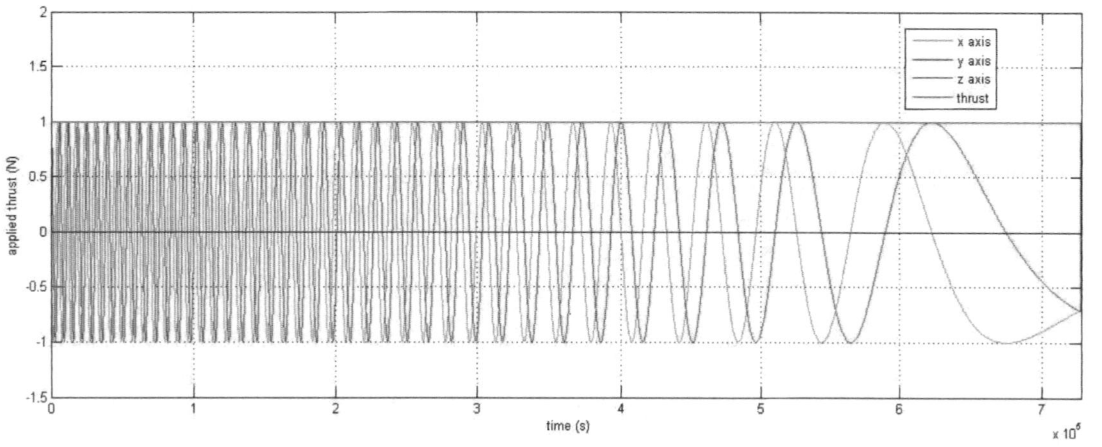

Fig. 3 –Thrust applied as a function of the time.

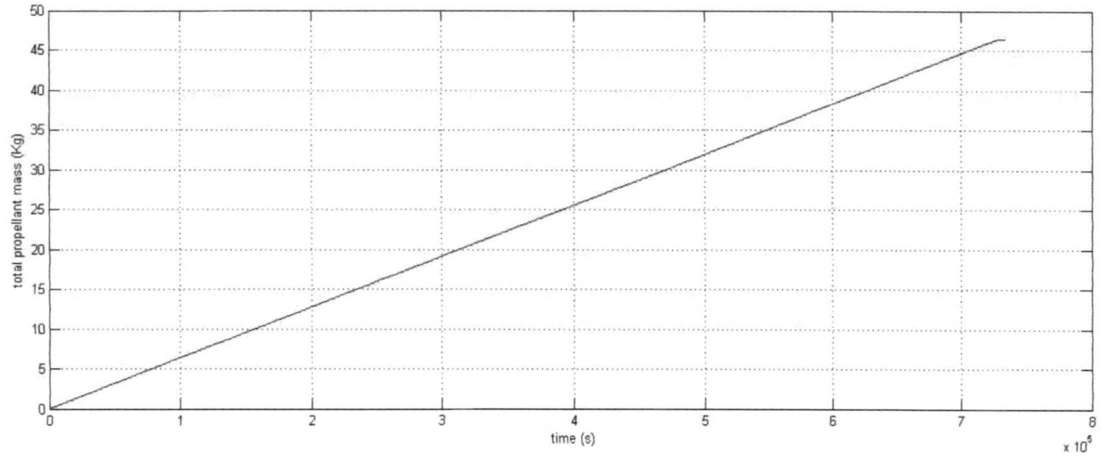

Fig. 4 – Fuel consumption as a function of the time.

6. Conclusions

This work studied the fuel consumption of maneuvers that send a spacecraft from a low altitude orbit around the Earth to an orbit that has apogee in the orbit of the Moon, as well as the time the spacecraft spend in the Van Allen Belts, for different kinds of propulsion systems.

The propulsion systems considered were two hall thrusters available under study at the Universidade de Brasilia. The study of orbital maneuvers considered several perturbation forces in the dynamical model, such that the J_2, J_3 and J_4 terms of the geopotential, the third-body perturbation due to the Sun and the Moon and the solar radiation pressure.

The model proposed for the Van Allen belts is simple, but provides a good estimative for the purposes of the present study. The belts are constantly changing and the density of the ionized particles varies constantly. In this way, the approach used here was to estimate an area for the Van Allen belts based on the average of the real data available.

The orbits with inclination equal to 90 degrees were the ones where the spacecraft spends less time inside the belts, because the geometry of these orbits makes the spacecraft to pass by latitudes below and above the limits of the belts, 65 degrees, so reducing the exposure of the spacecraft to the radiation. The 45 degrees of inclination and the equatorial orbits (zero inclination) have very similar results, since they both stay all the time inside the limits of the belts. The equatorial orbits implies in slight larger transit times, because these orbits are more perturbed and so it is necessary to apply the thrust for longer times in order to achieve the desired apogee distance.

The maneuvers to an orbit with the same apogee of the orbit of the Moon can have several purposes, such that sending a spacecraft to perform a Swing-By with the Moon to gain energy. This idea is being considered for the Aster mission, which is under study in Brazil, and has the goal of sending a spacecraft to an asteroid after gaining energy from the passage by the Moon.

Acknowledgment

The authors wish to express their appreciation for the support provided by grants N° 473387/2012-3 and 304700/2009-6, National Council for Scientific and Technological Development (CNPq); grants N° 2011/09310-7, 2011/08171-3 and 2011/24076-0, São Paulo Research Foundation (FAPESP).

References

[1] http://www.nasa.gov/mission_pages/rbsp/news/third-belt.

[2] Chen., F.F.,"Introduction to plasma physics and controlled fusion", Second Edition, Volume 1: Plasma Physics, Plenum Press.

[3] http://www.nasa.gov/mission_pages/rbsp/multimedia/20130228_briefing_materials.

[4] ROCCO, E. M. **Perturbed orbital motion with a PID control system for the trajectory**. In: COLÓQUIO BRASILEIRO DE DINÂMICA ORBITAL, 14., 2008, Águas de Lindóia. **Anais...** UNESP: Guaratinguetá, 2008, 2008a.

[5] FERREIRA, J. L.; POSSA G. C.; MORAES B. S.; SOUZA, J. H. C.; CARVALHO G. M ; MOURÃO, D.C.; WINTER O.C . Permanent magnet hall thruster for satellite orbit maneuvering with low power. **Advances in Space Research**, v. 01, p. 166-182, 2009. Série: 01. ISSN/ISBN(02731177).

[6] THAIS, C. O., ROCCO, E. M., FERREIRA J. L. F., PRADO, A. F. B. A., "Minimum Fuel Low-Thrust Transfers for Satellites Using a Permanent Magnet Hall Thruster," Mathematical Problems in Engineering, vol. 2013, Article ID 945030, 12 pages, 2013. doi:10.1155/2013/945030

[7] http://earthobservatory.nasa.gov/IOTD/view.php?id=78985

Journal of Physics: Conference Series **465** (2013) 012020 doi:10.1088/1742-6596/465/1/012020

A brief scenario about the "space pollution" around the Earth

T P Brito [1], C C Celestino [1], R V Moraes [2]

[1] Universidade Federal do ABC - UFABC, Av. dos Estados, 5001, 09210-588, Santo André, SP, Brasil

[2] Universidade Federal de São Paulo - UNIFESP, Rua Talim, 330, 12231-280, São José dos Campos, SP, Brasil

E-mail: tiago.brito@aluno.ufabc.edu.br

Abstract: In this work is presented a brief review about the main events generating of space debris around the Earth, occurred up to the present day. How the clouds of debris "polluted" the neighborhood of orbits in which the bodies were initially allocated is here analyzed. The implications of the growth of space debris existing on space missions as well as safety rules to control sources of debris are discussed.

Keywords: Space debris, historical scenario, main events generating debris

Introduction

In the second half of the twentieth century, more specifically at the post-Cold War, U.S. and the former USSR vied for supremacy between their forms of government by the called Space Race. This milestone effectively began with the launch by the Soviets of the artificial satellite called Sputnik 1 in October 1957. Since then, thousands of missions were carried out not only by the two superpowers, but also by several countries. After the Sputnik 1 approximately 28,000 objects were placed in orbit around the Earth, of which approximately 9,000 are still objects in orbit and of these only 6% still in operation [1]. At the end of its useful life, the vast majority of satellite remains in orbit around the planet. Added to inactive satellites there are still missions for which the final stages of the rocket launcher, adapters, and lens caps that remain in orbit beyond debris caused by explosions of satellites, or collisions between them, or even with other debris. All these objects should be characterized as space junk. Removal of some of these objects is made naturally by gravitational and non- gravitational perturbations It is estimate that the natural removal of objects smaller than 10 cm can last several days if these objects are below 200 km altitude, several years if their orbits are between about 200 and 600 km, hundreds of years for orbits between about 600 and 800 km, and permanently if the altitude is greater than about 36,000 km [1].

2 - Generating new debris

Explosions of some objects in orbit were detected and may be related to residual fuel remaining in tanks of stage of rocket launchers discarded around the Earth. Over time, the space environment may deteriorate the mechanical integrity of external and internal parts, leading to leaks or the mixture of fuel components, which can lead to auto ignition. The resulting explosion can destroy the source object and its mass is spread through numerous fragments with a wide range of masses and velocities transmitted [2]. Considering the events of collisions and

Journal of Physics: Conference Series **465** (2013) 012020 doi:10.1088/1742-6596/465/1/012020

explosions there are numerous records and so some of these may be described as major sources of generation of space debris. The first verified case of collision between two objects in space is assigned to the collision between the cataloged space debris coming from the rocket Ariane 1 with Cerise, a French military reconnaissance satellite. The crash occurred on July 24, 1996 and caused a cut of 4.3 meters in the system of stabilization by gravity in the Cerise, leaving the satellite severely damaged [3].

3 - The main events generating debris

Below will be mentioned some of the most significant events that have already occurred in **l**ow-altitude **e**arth **o**rbits - LEO, given in order of increasing of generation of cataloged objects.

Fengyun 1

In January 2007, the Fengyun 1C weather satellite that was out of operation was blown up by a Chinese ASAT (anti-satellite missile) as part of a widely criticized test of anti-satellites missiles. For a period of one month after the explosion, it is estimated that the number of fragments with sizes larger than 5 cm is above 2000 and up to 1 cm that number is about 35,000, [4]. After this incident it is estimated that the volume of debris in orbit of the planet has increased by 25%, from 9,949 cataloged objects in December 2006 to 16,094 in July 2011. The debris cloud generated after the collision is dispersed in orbits between 125 km altitude up to 3,800 km, affecting a significant portion of the LEO satellites [5].

The Chinese test conducted in January 2007 with anti-satellite missiles, which culminated in the explosion of the satellite Fengyun 1, represents the largest single event generator of space debris ever recorded, with approximately 3,037 objects with sizes equal to or greater than 10 cm, the Fengyun debris corresponds at 22% of the total volume of debris in orbit in the LEO region. The cloud of debris coming from the Fengyun 1C has an apogee range of about 400 km and 3600 km, and range for the perigee of about 350 km at 900 km altitude, with the region of highest density of this cloud corresponding to an altitude interval between 500 km and 1500 km. The debris from the Fengyun 1C cause risk not only to the International Space Station - ISS but also to the various satellites in LEO orbits. Studies done by NASA simulated the debris cloud evolution of the Fengyun 1C for the next 100 years since its fragmentation and the results is that, even for this long period, it is estimated that 79% of the Fengyun C debris still remain in orbit [6].

Iridium 33 e Cosmos 2251 - The collision between the U. S. Iridium-33 satellite, owned by the private company Iridium, and the Russian military satellite Cosmos is the second largest generator of waste already evident, and produced approximately 1,685 cataloged debris. This collision occurred on February 10/ 2009 about 800 km altitude, but their debris generated an immense cloud covering almost all orbits corresponding to the interval between 180 km and 1700 km altitude, which is a very dense region of active satellites orbiting between 600 km and 1000 km [7]. This event "polluted" almost every region LEO offering risk for any and every object that meets these orbits.

Breze-M - On 16 October 2012 the third stage of a Proton-M rocket exploded in orbit creating a cloud containing about 80 cataloged debris. This is not the third largest event causing debris, but has attracted more attention because the debris cloud generated covers orbits between 100 and 5500 km altitude, which endangers all satellites in LEO orbits and in particular shall ISS where the intersection of the two orbits are the debris which is the same altitude ISS [8].

There were also other major events generating debris in LEO orbits, like the explosion of the upper stage of the STEP II, Pegasus Propulsion System Assist the Hydrazine - HAPS, in 1996, where they were catalogued 702 debris that formed a cloud between approximately 350 km and 3000 km altitude endangering missions in low Earth orbit and will even International Space Station - ISS. And the fragmentation of Third Stage of launch vehicle Long March 4B occurred on March 11, 2000, generating a cloud of debris catalogued 345.

Figure 1 illustrates the evolution of the number of debris tracked by the Space Surveillance Network of the U.S. with emphasis on the rapid increase of the debris after the collision mentioned above. Looking at the chart it can be observed that the amount of debris orbiting the

Earth has more than doubled after the destruction of the Chinese satellite Fengyun and the collision between the satellites Iridium and Cosmos [9].

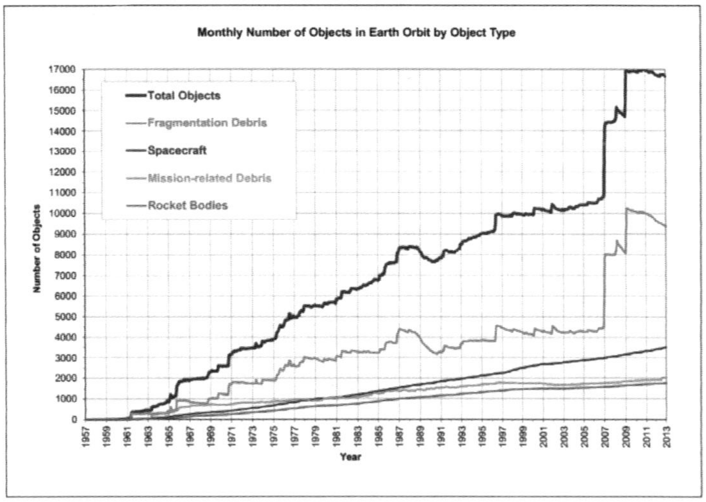

Figure 1 - Evolution of debris tracked by the Space Surveillance Network. (Extracted from [9]).

There were also other major events generating debris in LEO orbits, like the explosion of the upper stage of the STEP II, Pegasus Propulsion System Assist the Hydrazine - HAPS, in 1996, where they were catalogued 702 debris that formed a cloud between approximately 350 km and 3000 km altitude endangering missions in low Earth orbit and will even International Space Station - ISS. And the fragmentation of Third Stage of launch vehicle Long March 4B occurred on March 11, 2000, generating a cloud of debris catalogued 345.

3.1 Secondary events

Although not yet confirmed, there is strong evidence that two fragmentations resulted from collisions with debris from the Fengyun 1C.

By February 19, 2007 one of the injection stage of the Briz-M exploded on Western Australia. The cause of his outburst, though not yet demonstrated, is associated with debris from the collision will Fengyun 1C because it occurred about 37 minutes after its passage through the ring of debris generated by the event [10]. On February 14, 2007 the auxiliary engine of the SL-12 rocket exploded in orbit. This event occurred one month after the fragmentation of the Fengyun 1C. This event deserves greater attention and being technically examined, because it may have originated by a collision with debris from the Fengyun 1C [11].

All the cited events occurred in LEO orbits, but there are also events in other regions, **g**eostationary-altitude **e**arth **o**rbits (GEO) and **m**ean-altitude **e**arth **o**rbits (MEO). Table 1shows the apogee and perigee of the three main events in MEO and GEO regions.

MEO Events				GEO Events			
Object	Apogee	Perigee	Number of Debris	Object	Apogee	Perigee	Number of Debris
COSMOS 2022-2024	18.410	655	105	INTELSAT 515 R/B	35.720	510	37
COSMOS 1970-1972	18.515	720	76	COSMOS 1285	40.100	720	18
COSMOS 1883-1885	19.120	335	39	CAT R/B	33.140	180	18

Table 1 - Apogee, Perigee and number of debris to the main events in MEO and GEO regions.

4. Safety rules to control sources of debris

Nowadays it is estimated that there are about billions of space debris orbiting our planet with a diameter of less than 1 cm. There are about 100,000 debris orbiting around the Earth with diameter between 1cm and 10 cm and about 17,000 debris with a diameter greater than 10 centimeters [1]. The existence of debris orbiting Earth can endangers the safety of space missions and satellites. In order to limit the generation of orbital debris and considering the importance of this "pollution space" for technological advancement, NASA created the Safety Standard 1740.14 [12].

Among the conditions to be met by space missions, should also be performed:
• depleting energy sources on board after the completion of the mission to prevent the disintegration caused by bursts of residual fuel in the tanks and transmission lines;
• limit the life of the mission in orbit after 25 years performing maneuvers to a disposal orbit. These regions are most common for objects in GEO and MEO orbits and considering that the re-entry of objects at these altitudes is very unlikely; therefore, objects are allocated in storage orbits which serve as a kind of "graveyard" for satellites at the end of its useful life. The storage orbits are located down and up about 500 km with respect the orbits MEO and GEO, so this measure does not reduce the amount of "space junk" around the planet, only keeps the debris more far from the used orbits. For LEO orbits, due to its greater proximity from the Earth, the most interesting task to do at the end of the useful life of a satellite is providing its reentry "controlled" in the atmosphere, where due to the high-speed reentry friction with atmospheric air would cause fragmentation of the object. The most recent missions include in your planning an extra fuel to this maneuver. For altitudes up to 200 km reentry occurs naturally after a certain period, because in this region there is still a rarefied atmosphere which causes friction with the object. This atmospheric reentry "controlled" is made so that debris will fall into the ocean at least 46 km away from the mainland;
• limit the generation of debris associated with the normal spatial operations such as the introduction of extra fuel for the last stage rocket re-enter into the atmosphere;
• limit the risk of system components spatial to survive atmospheric reentry and thus prevent bodies from falling on populated areas endangering the planet's population.

5. Cloud of debris

Fragmentations and collisions of space objects give rise to a phenomenon known as "the clouds of debris "polluted" the neighborhood of orbits in which the bodies were initially allocated. This phenomenon may preclude executing future missions in orbits affected and endangers some specific missions. Currently there are no effective means capable of removing large amounts of space debris, and the growth in the number of debris in orbit of the planet has become a big problem. Thus, the following items will treat the evolution of debris clouds in different regions.
LEO orbits correspond to altitudes ranging from 200 to 2,000 km above the earth's surface. In these orbits satellites travel at a speed of about 27,400 km / h, this represents a revolution about 90 minutes. In this region the greatest concentration of debris space is found, the number of conjunctions which is proportional to the risk of collisions is about five kilometers to objects in LEO orbits in a period of 24 hours. After incidents involving the ASAT with the satellite Fengyun 1C in 2007 and the collision between the two satellites (Iridium and Cosmos) in 2009 greatly increased the number of debris and consequently the number of collisions in LEO orbits where are still found approximately 503 active satellites [13]. The Figure 2 shows the distribution of the population of debris in this region as a function of density, with emphasis on waste generated by the fragmentation of the satellite Fengyun 1C and by the collision between the satellites Iridium 33 and Cosmos 2251 [9].

Through time, various gravitational and non- gravitational forces will spread the particles beyond the original plane of the orbit. To illustrate, Figure 3 shows the behavior of the scattering of debris as function of the time. Note that, in just 80 days, the cloud of particles make up a ring around the Earth [9]. Figure 4 shows the four largest clouds of debris in LEO.

6. The estimated growth of space debris

A projection, for the growth of space debris for the next 200 years, taking into account the worst case, alert us to comply with the mitigation standards, without which, the space environment will become so polluted that will preclude future missions [15]. For such a projection is possible to notice that the LEO orbits require greater attention, because for these orbits is forecast exponential growth and this factor is known as the "Kessler Syndrome". This is what motivated the creation of the current mitigation measures, for example, a maximum of 25 years for a satellite to remain in orbit. After this period the satellite is relocated into a disposal orbit. For MEO and GEO orbits the scenario is a little more exciting because predicts moderate growth in waste generation for these orbits.

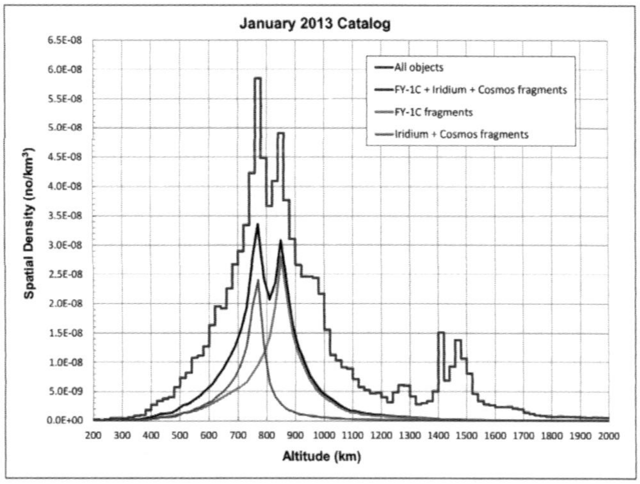

Figure 2 - Distribution of cataloged debris population in low Earth orbits before and after main events of explosion (Extracted from [9]).

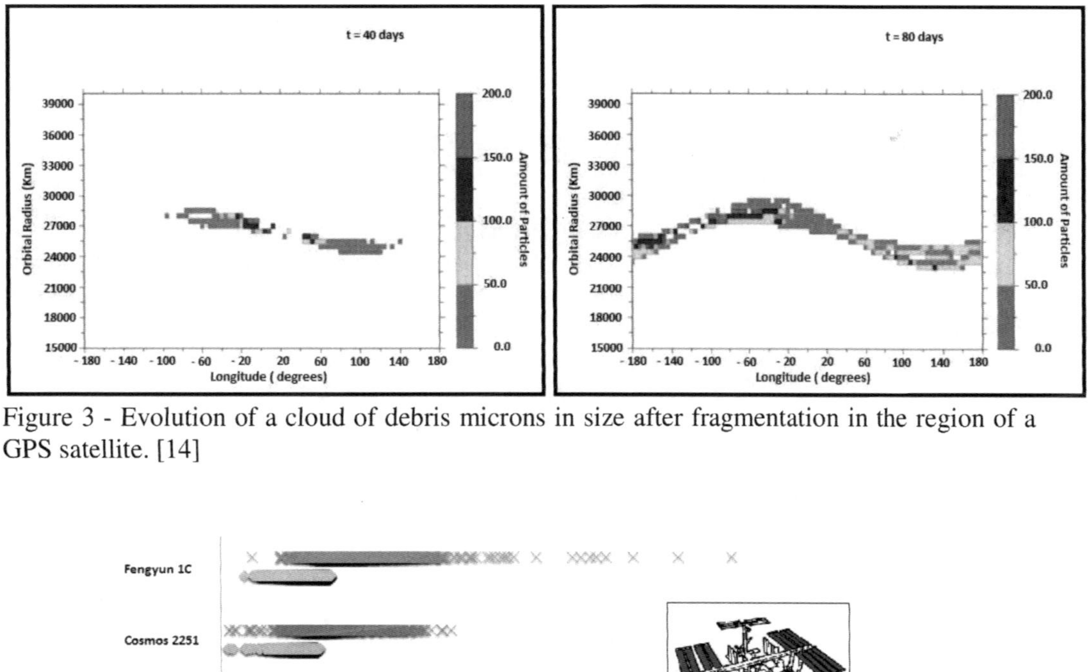

Figure 3 - Evolution of a cloud of debris microns in size after fragmentation in the region of a GPS satellite. [14]

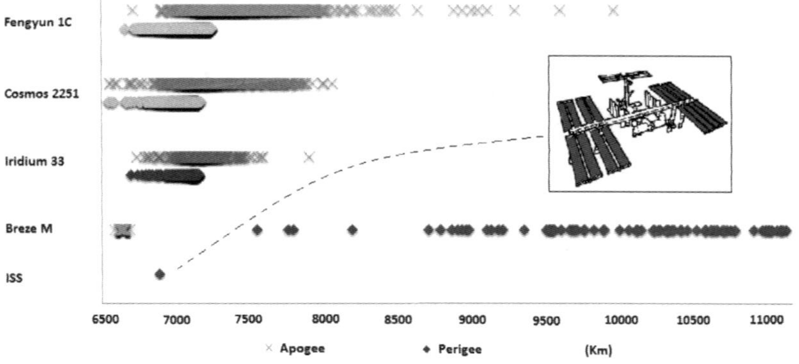

Figure 4: The four largest clouds of debris in LEO. The location of the ISS is shown.

Journal of Physics: Conference Series **465** (2013) 012020 doi:10.1088/1742-6596/465/1/012020

Conclusion

In this paper we presented a historical scenario of the environment around the Earth in terms of space debris. The main events considered sources of waste generation were cataloged and presented to estimate this future environment. These findings emphasize the importance of means of "cleansing" of regions around the Earth so that are not hindered future spatial missions and thus technological advances. Safety standards for the generation of space debris can ease the situation but not solve it fully. It would require a more effective surveillance so that these patterns could actually be met. Further research in parallel must be undertaken for more immediate and effective result. By this brief history it was observed that the formed cloud of space debris with less than 10cm may also offer lower risks but important since they are not cataloged. It is believed that this brief historical backdrop of space debris around Earth might contribute to future projects on this topic in Brazil.

References

[1] NASA. *Orbital Debris Education Package*. NASA Orbital Debris Program Office. 2012a. Available at: <http://www.orbitaldebris.jsc.nasa.gov/library/EducationPackage.pdf>. Accessed on Sep, 28, 2012.

[2] ESA. *Space Debris. History and background*. 2009. Available at: <http://www.esa.int/esaMI/Space_Debris/SEMQQ8VPXPF_1.html>. Accessed on: Sep. 27, 2012.

[3] NASA. *NSSDC Master Catalog Search - Cerise*. 2012b. Available at: <http://nssdc.gsfc.nasa.gov/nmc/spacecraftDisplay.do?id=1995-033B>. Accessed on: 30 Sep. 2012.

[4] NASA. *Orbital Debris Quarterly News Volume 11, Issue 2 April 2007*. Available at: <http://orbitaldebris.jsc.nasa.gov/newsletter/pdfs/ODQNv11i2.pdf>. Accessed on: 03 Oct. 2012.

[5] USSPACECOM. *Space Surveillance Network*. In 2012. Available at: <http://www.au.af.mil/au/awc/awcgate/usspc-fs/space.htm>. Accessed on 02 Oct. 2012.

[6] CELESTRAK. *Chinese ASAT Test*. Updated 2012. Available at:

[7] CELESTRAK. *Iridium 33/Cosmos 2251 Collision*. Updated 2012. Available at: <http://www.celestrak.com/events/collision/>. Accessed on: 29 Sep. 2012.

[8] CELESTRAK. *Breeze-M Rocket Body Breakup*. 2012. Available at: <http://www.celestrak.com/events/38746-Breakup/>. Accessed on: 17 Nov. 2012.

[9] NASA. *Orbital Debris Quarterly News Volume 17, Issue 1 January 2013*. Available at: <http://orbitaldebris.jsc.nasa.gov/newsletter/pdfs/ODQNv17i1.pdf>. Accessed on: 15 May. 2013.

[10] Carrico, Timothy; Carrico, John; Policastri, Lisa; Loucks, Mike. *Investigating orbital debris events using numerical methods with full force model orbit propagation*. AAS 2012.

[11] CELESTRAK. *Recents Debris Events. 2012*. Available at: <<http://www.celestrak.com/events/debris-events.asp >. Accessed on: 07 Nov. 2012.

[12] NASA. *Orbital Debris. Safety Standard 1740.14*. 2005. Available at: < http://orbitaldebris.jsc.nasa.gov/mitigate/safetystandard.html >. Accessed on: 30 Oct. 2012.

[13] David Wright. *The Current Debris Situation*. 2010. Available at: < http://swfound.org/media/99971/Wright-space-debris_situation.pdfl>. Accessed on: 02 Feb. 2013.

[14] Celestino, C. C. *Study of the Dynamics of Small Meteoroids and Debris Special*. Ph. D. Thesis, INPE, São José dos Campos, 2005.

[15] NASA. **Orbital** *Debris and Future Environment Remediation*. 2011. Available at: < http://www.nasa.gov/pdf/582393main_OCT-Orbital_Debris_TAGGED.pdf >. Accessed on: 25 Nov. 2012.

Journal of Physics: Conference Series **465** (2013) 012021 doi:10.1088/1742-6596/465/1/012021

Trajectory control with continuous thrust applied to a rendezvous maneuver

W G Santos and E M Rocco

National Institute for Space Research, Space Mechanics and Control Division, São José dos Campos, Brazil

E-mail: willer.gomes@dem.inpe.br

Abstract. A rendezvous mission can be divided into the following phases: launch, phasing, far range rendezvous, close range rendezvous and mating (docking or berthing). This paper aims to present a close range rendezvous with closed loop controlled straight line trajectory. The approaching is executed on V-bar axis. A PID controller and continuous thrust are used to eliminate the residual errors in the trajectory. A comparative study about the linear and nonlinear dynamics is performed and the results showed that the linear equations become inaccurate insofar as the chaser moves away from the target.

1. Introduction

Rendezvous and docking or berthing (RVD/B) technology are key elements in mission such as assembly in orbit of larger units, re-supply of orbital platforms and stations, exchange of crew in orbital stations, repair of spacecraft in orbit, retrieval of spacecraft, and others. The RVD/B process includes several orbital maneuvers and trajectory control. On 16 March 1966, took place the first rendezvous and docking between two spacecrafts, and the first automatic RVD took place only on 30 October 1967. A rendezvous mission can be divided into a number of phases: launch, phasing, far range rendezvous, close range rendezvous and mating.

This paper will focus on close range rendezvous phase, which can be divided into two sub-phases: a preparatory phase leading to the final approach corridor, often called closing, and a final approach phase leading to the mating conditions. The objectives of the closing phase are the reduction of the range to the target and the achievement of conditions allowing the acquisition of the final approach corridor. Different acquisition strategies for V-bar and R-bar approaches can be executed. The objective of the final approach phase is to achieve docking or berthing capture conditions in terms of positions and velocities and of relative attitude and angular rates. The straight line or the quasi-straight line trajectory are used in this maneuver phase. The guidance, navigation and control (GNC) system of the chaser must following the direction of the target docking axes. To meet this requirement, the relative attitude between the docking ports of the two spacecrafts should be measured and taken to zero at the end of the maneuver. For observability and safe reasons a cone-shaped approach corridor will usually be defined, within which the approach trajectory has to remain. This corridor has a half cone angle of 10-15 deg [1].

A comparison between a linear dynamics model - take into account a circular orbit - and a nonlinear dynamics model - for an orbit with eccentricity of 0.1 - was done by [2]. Although his main intention was present a new method for applying the Hill equations to the elliptical orbits, this comparative study is a valuable reference. So, the purposes of this paper are the following: analyze the relative motion for several initial conditions; find out the difference between the linear and nonlinear spacecraft dynamics considering the same orbital eccentricity; and to perform a close range rendezvous with closed loop controlled straight line trajectory on V-bar axis. The Section 2 shows the equations of the control system. The simulations and results are shown in Section 3. Lastly, a conclusion is given in Section 4.

2. Mathematical modeling

To perform a close range rendezvous it was developed a trajectory control system. The control system has the function of correcting the errors in the chaser trajectory with respect to a reference trajectory. It was adopted a PID controller and continuous thruster as actuator. The basic diagram of the closed loop control system is presented in Figure 1.

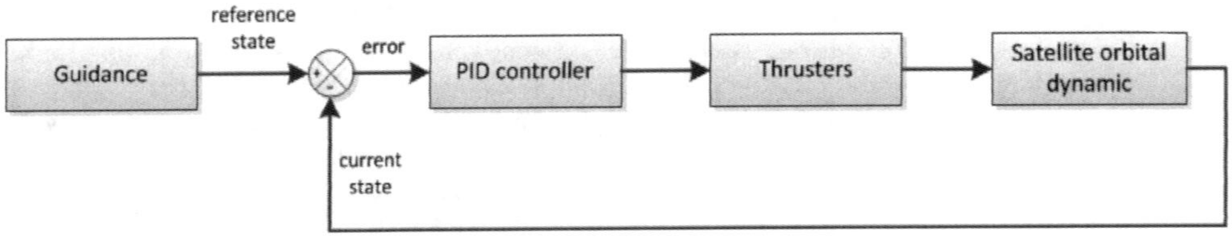

Figure 1. Block diagram of the control system.

It was considered a constant approaching velocity for calculation of the reference trajectory. The **Guidance** function is given by,

$$x_r(t) = x_0 - v_x t \tag{1}$$

where x_r is the reference position; x_0 is the initial distance of the chaser from the target; v_x is the approaching velocity on the V-bar axis; and t is the time.

Several types of controllers can be used in a control system, according to their application, such as nonlinear controller, LQR, LQG, PID, PID-fuzzy, H_∞, etc. Most industrial controllers are Proportional-Integral-Derivative (PID) due to its flexibility, low cost and robustness. For applications in RVD/B maneuvers, **PID controllers** usually meets the needs and its control law is given by [3]

$$c(t) = K_p e(t) + K_i \int e(t)dt + K_d \frac{de(t)}{dt} \tag{2}$$

where $e(t)$ is the state error; K_p, K_i and K_d are the proportional, integral and derivative gains, respectively.

Actuators are devices that operate on the spacecraft in order to maintain its orbit and attitude or change their orientation and orbit through maneuvers. **Thrusters** are the only actuators that can be used both to orbital maneuvers as for attitude maneuvers, but the thrust magnitude is different for each purpose. Spacecraft propulsion systems are divided into three categories: cold gas, chemical (solid and liquid), and electrical. The basic equation of propulsion holds for all kinds of propellants. A rocket engine develops its thrust (F) by expelling propellant (such as gas molecular or ions) at a high exhaust velocity (V_e) relative to the satellite body. The amount of thrust can be calculated as follow [4],

$$F = V_e \frac{dm}{dt} + A_e[P_e - P_a] = V_{ef} \frac{dm}{dt} \tag{3}$$

where P_e and P_a are the gas and ambient pressures, respectively; V_e is the exhaust velocity; V_{ef} is the effective exhaust velocity of the expelled mass with respect to the satellite; dm/dt is the mass flow rate of the propellant; and A_e denotes the area of the nozzle exit.

During the close range rendezvous the chaser is a few tens of meters from the target and therewith the relative navigation assumes greater importance. The following considerations are taken in the development of equations: the spacecraft is treated as a mass point; the chaser motion is disturbed by the effects of spherical earths gravitational fields and the thruster forces. Consider **X**, **Y** and **Z** as axis of the inertial frame; **x**, **y** and **z** are the axis of the target centered local orbital frame; $\mathbf{r}_c$ and $\mathbf{r}_t$ is vector radius of the chaser and target, respectively, as shown in Figure 2. In the rendezvous literature, the **x** axis, **y** axis and **z** axis is called by V-bar, H-bar and R-bar, respectively.

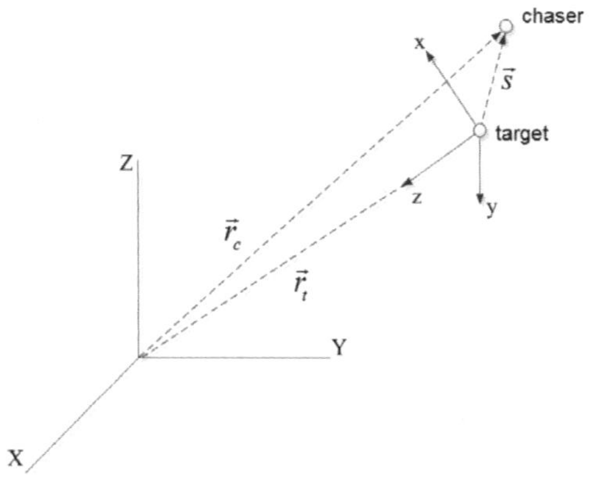

Figure 2. Local orbital frame and inertial frame.

The relative acceleration $\ddot{\mathbf{s}}$ is

$$\ddot{\mathbf{s}} = \ddot{\mathbf{r}}_c - \ddot{\mathbf{r}}_t \tag{4}$$

The general equations for motion under the influence of a central force are written as [1]

$$\begin{aligned} \ddot{\mathbf{r}}_t &= -\mu \frac{\mathbf{r}_t}{r_t^3} \\ \ddot{\mathbf{r}}_c &= -\mu \frac{\mathbf{r}_c}{r_c^3} + \mathbf{f_u} \end{aligned} \tag{5}$$

where μ is the earth's gravitational constant and $\mathbf{f_u}$ is the specific force vector of control. The equations 4 and 5 represent the **nonlinear dynamics** of the system.

The target centered local orbital frame is a non-inertial frame and it is rotating around the central body with an orbital rate ω. It is common to express the relative motion by linear time varying differential equations, as shown in [1],

Journal of Physics: Conference Series **465** (2013) 012021 doi:10.1088/1742-6596/465/1/012021

$$\begin{aligned} \ddot{x} - 2\omega\dot{z} &= f_x \\ \ddot{y} + \omega^2 y &= f_y \\ \ddot{z} + 2\omega\dot{x} - 3\omega^2 z &= f_z \end{aligned} \tag{6}$$

where ω is the orbital angular velocity that for the special case of circular orbits is considered constant and can be expressed by,

$$\omega^2 = \frac{\mu}{r_t^3} \tag{7}$$

and f_x, f_y and f_z are the components of the specific force vector of control.

The Equations 6 represent the **linear dynamics** of the relative motion and are known as **Hill equations**. The derivation of these equations starts from the linearization of the nonlinear equations and can be consulted in [1] and [5].

3. Results

This section will present the results about the relative motion between two spacecrafts. Two study cases were prepared: the first shows the relative motion dynamic for different initial conditions of the chaser vehicle, a comparison of the nonlinear and linear dynamics is performed in order to identify the error originating from the linearization and the second case illustrates a rendezvous maneuver considering closed loop controlled straight line trajectory. All results are presented in the target centered local orbital frame.

It was used a simulation step of 1 second. The initial keplerian elements of the target vehicle are shown in Table 1.

Table 1. Initial conditions of the target.

Parameter	Value
Altitude (km)	400
Eccentricity	0
Inclination (degrees)	30
RAAN (degrees)	0
Perigee argument (degrees)	0
Mean anomaly (degrees)	10

3.1. Case 1: Analysis of the relative motion dynamics

In this first case, the relative motion of the linear and nonlinear dynamics was evaluated for two orbits of the target. The initial condition of the chaser is shown in the legend of each figure. The first situation is considered with the chaser starting at 10 m away from the target in the direction of the R-bar axis, the relative motion can be seen in the Figure 3. It can be observed that after two orbits, the chaser is approximately 750 m away from the target. In order to evaluate the error from the linearization, the Figure 4 presents the difference between the models, considering the initial conditions of the Figure 3, as a function of target distance. Exceptionally in this graph, the stopping criterion was changed to four orbits of the target. It

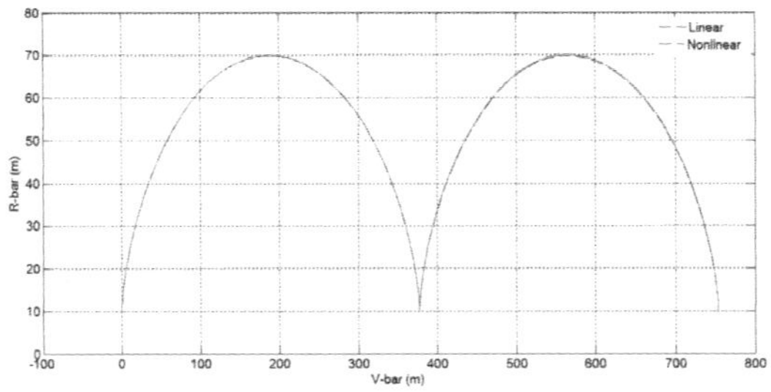

Figure 3. Case 1: $x_0, y_0 = 0$; $z_0 = 10$ m; $\dot{x_0}, \dot{y_0}, \dot{z_0} = 0$.

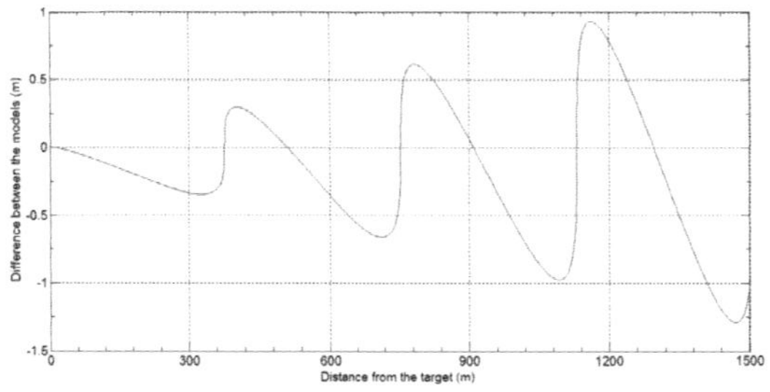

Figure 4. Case 1 (difference between the models): $x_0, y_0 = 0$; $z_0 = 10$ m; $\dot{x_0}, \dot{y_0}, \dot{z_0} = 0$.

is noted that the Hill equations lose their accuracy since the distance between the spacecrafts increases, reaching an error of the order of 1 m in 1500 km from the target.

Consider that the spacecrafts are coupled and the chaser is subjected to a impulse of magnitude of 0.01 m/s in the direction of +V-bar, ie, a impulsive ΔV in the tangential direction to the orbit. The resulting movement is depicted in Figure 5. Now the same impulse is applied, however, in the radial direction to the orbit, ie, toward +R-bar. The relative dynamic behavior, in this case, can be seen in Figure 6. The resulting trajectory is an ellipse, that is, the spacecraft returns to the initial position each orbital revolution. From the Figures 5 and 6 we can conclude that the displacements due to a radial ΔV are much smaller than those due to a tangential ΔV of the same size.

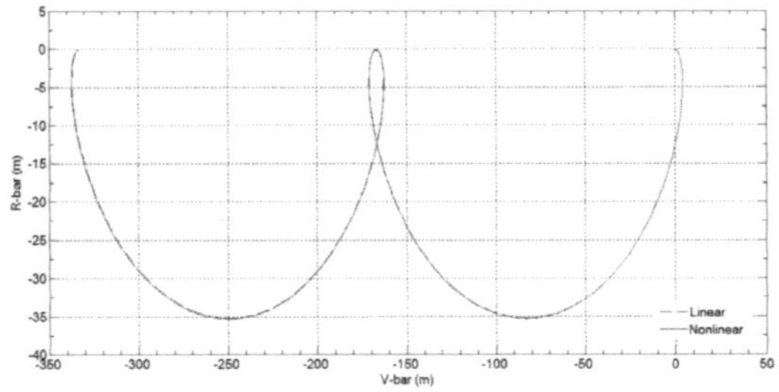

Figure 5. Case 1: $x_0, y_0, z_0 = 0$; $\dot{x_0} = 0.01$ m/s; $\dot{y_0}, \dot{z_0} = 0$.

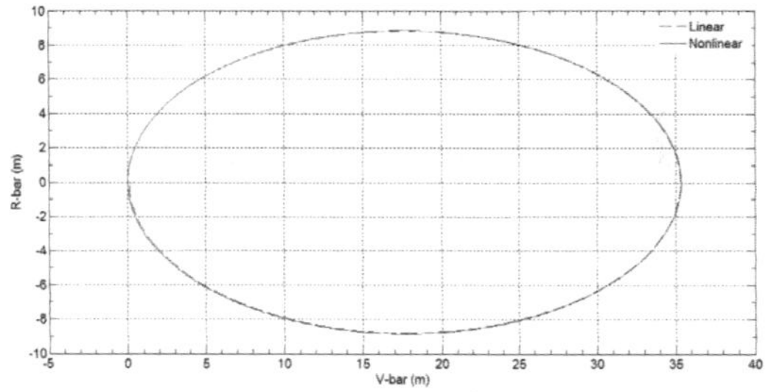

Figure 6. Case 1: $x_0, y_0, z_0 = 0$; $\dot{z_0} = 0.01$ m/s; $\dot{x_0}, \dot{y_0} = 0$.

The relative motion, considering the initial condition of 10 m in the H-bar axis, as a function of time, is presented in the Figure 7. We can highlight the sinusoidal behavior of the motion when the vehicle starts from a point outside the orbital plane.

3.2. Case 2: Close range rendezvous with closed loop controlled straight line trajectory

In this second case, it was considered a trajectory control system to perform a close range rendezvous in the V-bar axis. The chaser is 100 m from the target in the V-bar direction and it is wanted to perform the rendezvous with a constant velocity of 0.05 m/s. The Figure 8 shows the transitory response with the linear and nonlinear dynamics. The movement in each axis as function of time is presented in Figure 9. It is observed that in the V-bar axis, the vehicle has met the target at a constant velocity, as desired; movement on the H-bar axis is not occurred because there was not application of forces outside the orbital plane, and there was a deviation in the R-bar axis due to coupling between the R-bar and V-bar axis.

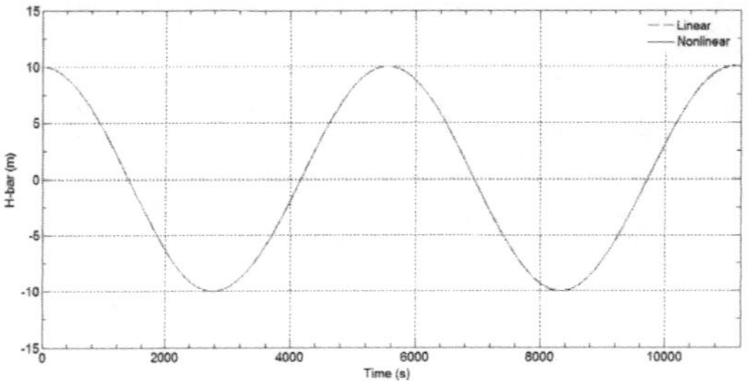

Figure 7. Case 1: $x_0, z_0 = 0$; $y_0 = 10$ m; $\dot{x_0}, \dot{y_0}, \dot{z_0} = 0$.

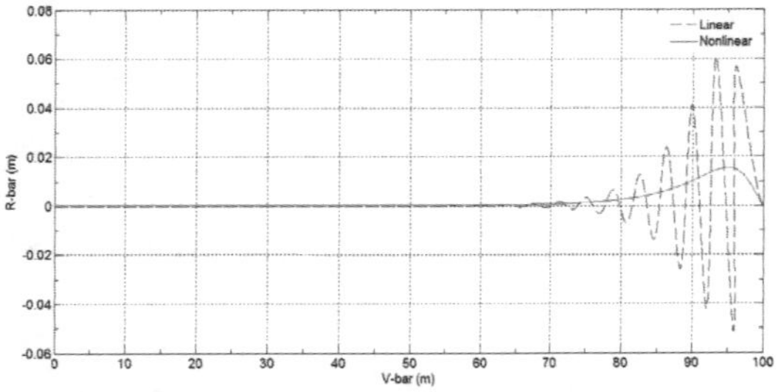

Figure 8. Case 2 (R-bar $\times$ V-bar): $x_0 = 100$ m; $y_0, z_0 = 0$; $\dot{x_0}, \dot{y_0}, \dot{z_0} = 0$; $v_x = 0.05$ m/s.

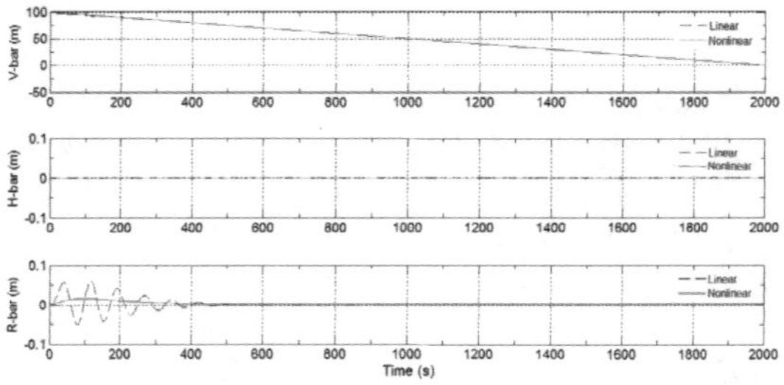

Figure 9. Case 2 (separated by axes): $x_0 = 100$ m; $y_0, z_0 = 0$; $\dot{x_0}, \dot{y_0}, \dot{z_0} = 0$; $v_x = 0.05$ m/s.

4. Conclusions

This paper presented a study about rendezvous maneuver. In the first case, free drift motion and impulsive maneuvers were analyzed for several initial conditions. The divergence between the linear and nonlinear spacecraft dynamics were also investigated. The second case showed the application of the trajectory control system in a close range rendezvous. The first results indicated that an specific motion can be yielded for each initial condition, like a sinusoidal, radial or cyclic behavior. Because of the linearization, the Hill equations become inaccurate insofar as the chaser moves away from the target. So, for long distances the nonlinear model is most advisable. Although we have different transitory responses, the trajectory control system met the requirements and executed the close range rendezvous with both models.

References

[1] Fehse W 2003 *Automated rendezvous and docking of spacecraft* (New York, USA: Cambridge University Press) 495

[2] Yamanaka K and Ankersen F 2002 New state transition matrix for relative motion on an arbitrary elliptical orbit *J. Guid. Control Dynam.* **25** 60-66

[3] Ogata K 1970 *Modern control engineering* (Englewood Cliffs, NJ: Prentice-Hall) 836

[4] Sidi M J 1997 *Spacecraft dynamics and control: A practical engineering approach* (New York, USA: Cambridge University Press) 409

[5] Arantes Jr G and Rocco E M and Komanduri A S 2010 *Far and close approaching strategies for rendezvous and docking operations applied to on-orbit servicing* (Mathematical analysis and applications in engineering aerospace and sciences vol 5) ed S Sivasundaram (USA: Cambridge Scientific) pp 1-29

Journal of Physics: Conference Series **465** (2013) 012022 doi:10.1088/1742-6596/465/1/012022

Attitude stability analyses for small artificial satellites

W R Silva[1], M C Zanardi[2], J K S Formiga[3], R E S Cabette[4] and T. J. Stuchi[5]

[1]Space Mechanics and Control Division, National Institute for Space Research (INPE), São Paulo, São José dos Campos, 12227-010, Brazil
[2]Department of Mathematics, Group of Orbital Dynamics and Planetology, São Paulo State University (UNESP), São Paulo, Guaratinguetá 12516-410, Brazil
[3]Faculty of Tecnology (FATEC), São Paulo, São José dos Campos, 12248-004, Brazil
[4]São Paulo Salesian University (UNISAL), São Paulo, Lorena, 12600-100, Brazil
[5]Rio de Janeiro Federal University (UFRJ), Rio de Janeiro,Rio de Janeiro, 21941-010, Brazil.

E-mail: reis.william@gmail.com

Abstract. The objective of this paper is to analyze the stability of the rotational motion of a symmetrical spacecraft, in a circular orbit. The equilibrium points and regions of stability are established when components of the gravity gradient torque acting on the spacecraft are included in the equations of rotational motion, which are described by the Andoyer's variables. The nonlinear stability of the equilibrium points of the rotational motion is analysed here by the Kovalev-Savchenko theorem. With the application of the Kovalev-Savchenko theorem, it is possible to verify if they remain stable under the influence of the terms of higher order of the normal Hamiltonian. In this paper, numerical simulations are made for a small hypothetical artificial satellite. Several stable equilibrium points were determined and regions around these points have been established by variations in the orbital inclination and in the spacecraft principal moment of inertia. The present analysis can directly contribute in the maintenance of the spacecraft's attitude.

1. Introduction

This work aims at analyzing the stability of the rotational motion of artificial satellites in circular orbit with the influence of gravity gradient torque, using the Andoyer's canonical variables. This stability analysis is very important in maintaining the attitude to ensure the success of a space mission. In this paper, Kovalev-Savchenko theorem (KST) [1] is used for the study of the stability and it ensures that the motion is Liapunov stable.

The objective of this paper is to optimize the stability analysis developed in [2,3] for the satellite in a circular orbit by applying the analytical expressions obtained in [4,5] for the coefficients of the normal 4th order Hamiltonian. The equilibrium points are established when terms associated with gravity gradient torque acting on the satellite are included in the equations of rotational motion. The stables regions around the equilibrium points are obtained by the variations in the orbital inclination and in the principal moment of inertia of the satellite.

In order to simplify the application of methods of stability in this is study, the Andoyer's variables [3,6] are used to describe the rotational motion of the satellite. These variables are represented by generalized moments (L_1, L_2, L_3) and by generalized coordinates (ℓ_1, ℓ_2, ℓ_3) that are outlined in Figure 1. The angular variables ℓ_1, ℓ_2, ℓ_3 are angles related to the satellite system Oxyz (with axes parallel of the spacecraft's principal axes of inertia) and equatorial system OXYZ (with axes parallel to the axis of the Earth's equatorial system). Variables metrics L_1 , L_2 , L_3 are defined as: L_2 is the magnitude of the angular momentum of rotation $\overrightarrow{L_2}$, L_1 and L_3 are, respectively, the projection of $\overrightarrow{L_2}$ z-axis's principal axis system of inertia ($L_1 = L_2 \cos J_2$, where J_2 are the angle between the z-satellite axis and $\overrightarrow{L_2}$) and projection $\overrightarrow{L_2}$ on the Z-equatorial axis ($L_3 = L_2 \cos I_2$, where I_2 are the angle between Z-equatorial axis and $\overrightarrow{L_2}$).

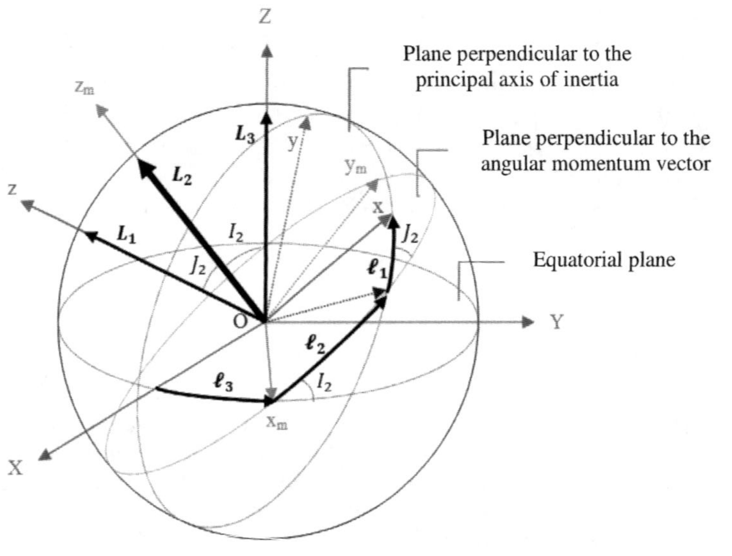

Figure 1. Andoyer' canonical variables [9]

In this paper, numerical simulations are made for a small artificial satellite which has orbital data and physical characteristics similar to real satellite. The stables regions around the equilibrium points are obtained by the variations of the orbital inclination and variations in the principal moments of inertia of the satellite.

2. Equation of motion

Andoyer's variables $(L_1, L_2, L_3, \ell_1, \ell_2, \ell_3)$, are used to characterize the rotational motion of a satellite around its center of mass [6] and the Delaunay's variables describe the translational motion of the center of mass of the satellite around the Earth [6].

Thus, assuming that the satellites have well-defined circular orbit, the goal is to study the stability of the rotational motion of the satellite. Then the Hamiltonian of the problem is expressed in terms of the Andoyer and Delaunay variables (L, G, H, l, g, h), including the gravity gradient torque, as follows [6,7]:

$$F(L_1, L_2, L_3, \ell_2, \ell_3, L, G, H, l, g, h) = F_o(L, L_1, L_2) + F_1(L_1, L_2, L_3, \ell_2, \ell_3, L, G, H, l, g, h) \quad (1)$$

with

$$F_o = -\frac{\mu^2 M^3}{2L^2} + \frac{1}{2}\left(\left(\frac{1}{C} - \frac{1}{2A} - \frac{1}{2B}\right) {L_1}^2 + \frac{1}{2}\left(\frac{1}{A} + \frac{1}{B}\right) {L_2}^2\right) + \frac{1}{4}\left(\frac{1}{B} - \frac{1}{A}\right)\left({L_2}^2 - {L_1}^2\right) \cos 2\ell_1 \quad (2)$$

$$F_1 = \frac{\mu^4 M^6}{L^6}\left[\frac{2C-A-B}{2} \hbar_1(\ell_m, L_n) + \frac{A-B}{4} \hbar_2(\ell_m, L_n)\right] \quad (3)$$

where $m = 2,3$ and $n = 1,2,3$; A, B and C are the principal moments of inertia of the satellite on x-axis, y-axis and z-axis respectively ; $\hbar_1$ and $\hbar_2$ are functions of the variables (ℓ_m, L_n), where ℓ_2 and ℓ_3 appear in the arguments of cosines. The complete analytical expression for perturbed Hamiltonian F_I is presented in [7] for circular orbit.

The equations of motion associated with the Hamiltonian F, Eq. 1, are given by:

$$\begin{cases} \frac{d\ell_i}{dt} = \frac{\partial F}{\partial L_i}, \\ \frac{dL_i}{dt} = -\frac{\partial F}{\partial \ell_i}; \end{cases} \qquad (i = 1,2,3). \tag{4}$$

These equations are used to find the possible equilibrium points of the rotational motion when will be considered two of its principal moments of inertia equal, $B = A$ (symmetrical satellite). With this relationship, the variable ℓ_1 will not be present in the Hamiltonian, reducing the dynamic system to two degrees of freedom, a necessary condition for applying the stability theorem chosen for analysis of equilibrium points.

3. The Algorithm for Stability Analysis

To use the *KST* is necessary the normal Hamiltonian of the problem. It was discussed in [3,8] and the normal Hamiltonian $\mathcal{H}$ is an analytic function of generalized coordinates (q_v) and moments (p_v) to a fixed point P, expressed by [3,8]:

$$\mathcal{H} = \sum_{v=1}^{2} \frac{\omega_v}{2} \mathcal{R}_v + \sum_{v,\upsilon=1}^{2} \frac{\delta_{v,\upsilon}}{4} \mathcal{R}_v \mathcal{R}_\upsilon + O_5 \tag{5}$$

where O_5 represents higher order terms; ω_v is the imaginary part of eigenvalues associated with the matrix defined by the product of a 4th order matrix symplectic with the Hessian of the Hamiltonian expanded in Taylor series up to 2nd order around the equilibrium point; $\delta_{v,\upsilon}$ depend on the eigenvalues ω_v and the coefficients of the Hamiltonian expanded in Taylor series of 3rd and 4th order around the equilibrium point, which are presented analytically in Formiga [5]; and

$$\mathcal{R}_m = q_m{}^2 + p_m{}^2 \qquad m = 1,2 \tag{6}$$

The stability analysis is performed here by the theorem Kovalev and Savchenko [1] which ensures that the motion is Liapunov stable if the following conditions are satisfied:

i. The eigenvalues of the reduced linear system are pure imaginary $\pm i\omega_1$ e $\pm i\omega_2$;

ii. The condition

$$k_1\omega_1 + k_2\omega_2 \neq 0 \tag{7}$$

is valid for all k_1 and k_2 integer satisfying the inequality.

$$|k_1| + |k_2| \leq 4 \tag{8}$$

iii. The determinant D must satisfy the inequality

$$D = -(\delta_{11}\omega_2{}^2 - 2\delta_{12}\omega_1\omega_2 + \delta_{22}\omega_1{}^2) \neq 0 \tag{9}$$

where δ_{uv} are the coefficients of the normal 4th order Hamiltonian.

4. Numerical simulations

In order to do the numerical simulations, in this paper were considered a small satellite (*SS*), which has similar orbital characteristics of the Second Brazilian Data Collection Satellite SCD-2 [2]. All the numerical simulations were developed using the software MATHEMATICA. The initial data for the SS satellite are (Table 1):

Table 1. Orbital characteristics of the Second Brazilian Data Collection Satellite SCD-2.

Orbit sets	Inclination	$I = 0.4364\, rad$
	Eccentricity	$e = 0$
	Orbital radius	$r = 7139.61585 km$
Principal moments of inertia on	x-axis	$A = 9.855 * 10^{-6} kg\ km^2$
	y-axis	$B = 9.855 * 10^{-6} kg\ km^2$
	z-axis	$C = 13.00 * 10^{-6} kg\ km^2$
Delaunay Variables	generalized moments	$L = 5334653.709\ kg\ km^2/s$
		$G = 5334653.709\ kg\ km^2/s$
		$H = 4834685.585\ kg\ km^2/s$
	generalized coordinates	$l = 0\, rad$
		$g = 4.5420\, rad$
		$h = 4.542\, rad$

There were 50 equilibrium points and only 7 were stables, the others 43 equilibrium points had also failed in the first condition of the *KST* and were not linearly stable, more details are presented in [9].

4.1. *Stables region around the equilibrium points*

The behavior around the Lyapunov stable equilibrium points were obtained by the variations in the orbital inclination (I) and in the principal moments of inertia of the satellite on x – axis (A) and on z - axis (C). For this procedure, was used the algorithm presented in [9].

Figure 2 shows the linear stabilization when the principal moment of inertia on the x-axis is fixed ($A_0 = 9.855 * 10^{-6} kg\ km^2$). The blue colour, in the region 1 and 2, represents the situation where the eigenvalues are pure imaginary. By the results it is possible to note that there are non-linear stabilization regions when $C = A_0$ or for low orbital inclination and there is linear stabilization regions for $C < A_0$ and $C > A_0$.

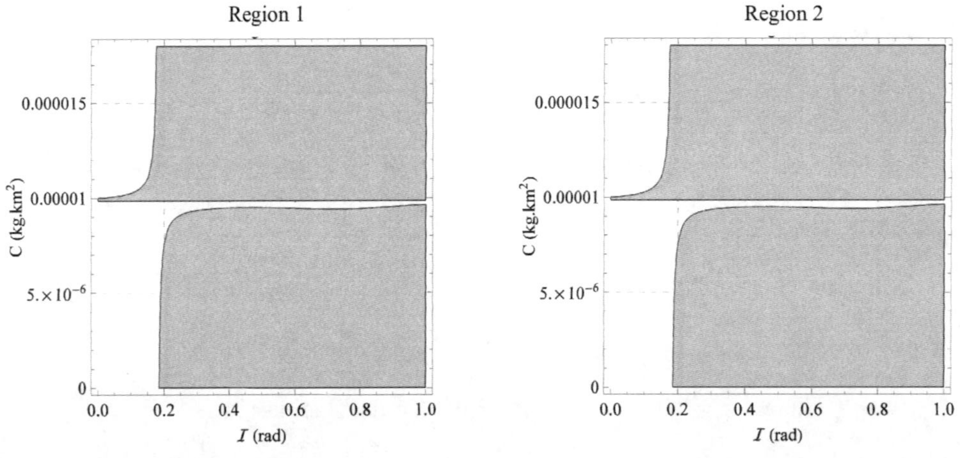

Figure 2. Linear stability region when $A_0 = 3{,}9499 * 10^{-1} kg\ km^2$

Figure 3 shows the analysis of the second condition for the *KST*, for the case when the principal moment of inertia on x –axis and the orbital inclination are fixed $A_0 = 9.855 * 10^{-6} kg\ km^2$ and $C_0 = 13 * 10^{-6} kg\ km^2$.

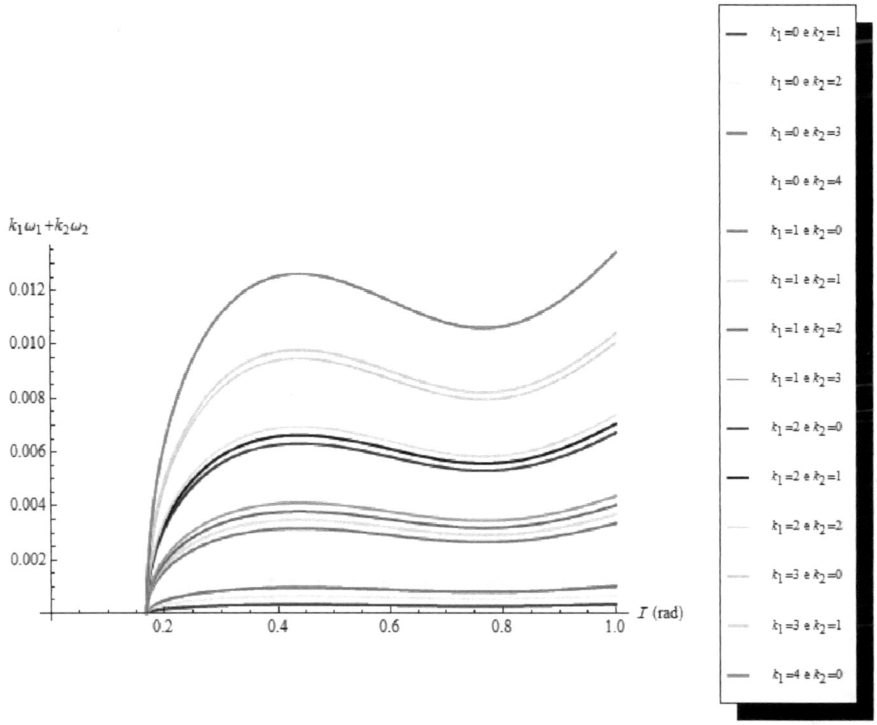

Figure 3 – Second condition for $k_1 \geq 0$ and $k_2 \geq 0$, considering $A_0 = 9.855 * 10^{-6} kg\ km^2$ and $C_0 = 13 * 10^{-6} kg\ km^2$.

Then the second condition ($k_1\omega_1 + k_2\omega_2 \neq 0$) is satisfied for $I > 0.1651\ rad$, in according with the results presented in the Figure 2.

By Figure 4, when the I_0 and A_0 are fixed, it is possible to see that the Arnold determinant is zero for values of the principal moment of inertia on z-axis (C), $9.5 * 10^{-6} kg\ km^2 < C < A_0$, than the third conditions are not satisfied and the rotational motion is nonlinear unstable.

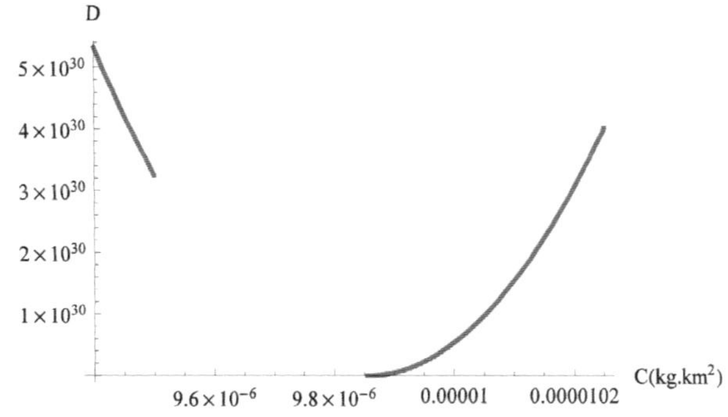

Figure 5. Arnold determinant in terms of orbital inclination, $A_0 = 9{,}855 * 10^{-6}\ kg\ km^2$ and $I_0 = 0.4364\ rad$

Journal of Physics: Conference Series **465** (2013) 012022 doi:10.1088/1742-6596/465/1/012022

5. Conclusions

In this paper it was presented a semi-analytical stability of the rotational motion of artificial satellites, considering the influence of gravity gradient torque for symmetric satellite in a circular orbit. Applications were made for small satellites (*SS*). Initially the points of equilibrium were determined using the physical, orbital and attitude characteristics of the satellite. Then the algorithm for stability analysis was applied and it was obtained 7 stable points for the SS satellite.

Several stable equilibrium points were determined and regions around these points have been established by variations in the orbital inclination and in the spacecraft principal moment of inertia. There were found 50 equilibrium points for the small size satellite (with some data similar to Second Brazilian Data Collecting Satellite) with 10 Liapunov stable points.

The results for the stable regions show that in the linear stability there is a separation between the stable and unstable region when the spacecraft principal moments of inertia are equals. It is also possible to observe that the rotational motion for the small satellite is linearly unstable in a low orbital inclination. For considered equilibrium points, the second condition is valid for all values of k_1 and k_2, for the small satellite it is necessary an orbital inclination bigger than 9.45 $degrees$. In the nonlinear analysis it was possible to verify that the linear stability doesn't guaranty the non-linear stability and the stable regions are bigger for the small satellite. For the considered satellite there is nonlinear stability for orbital inclination bigger than 9.45 $degrees$.

Then the present analysis can directly contribute in the maintenance of the spacecraft's attitude. Once the regions of stability are known for the rotational motion, a smaller number of maneuvers to maintain the desired attitude can be accomplished. In this case, a fuel economy can be generated to the satellite with propulsion systems control, increasing the spacecraft' lifetime.

Acknowledgments
The authors would like to thank the financial support received by CAPES and FAPESP (Proc. N°. 2012/21023-6).

References

[1] Kovalev, A. M. and A. Ia. Savchenko, "Stability of uniform rotations of a rigid body about a principal axis". PMM, v. 39, n. 4, 650 – 660, 1975.

[2] Cabette, R. E. S. "Estabilidade do movimento rotacional de satélites artificiais". Phd Thesis, National Institute for Space Research, São José dos Campos, Brazil, 2006.

[3] Vilhena de Moraes, R., Cabette, R. E. S., Zanardi, M. C., Stuchi, T. J. and Formiga, J K. "Attitude stability of artificial satellites subject to gravity gradient torque". Celes. Mech., 104, 337 – 353, 2009.

[4] Machuy, A. L. F."Cálculo efetivo da forma normal parcial para o problema de Hill". 71p. Dissertation, Instituto de matemática, UFRJ - Universidade Federal do Rio de Janeiro, 2001.

[5] Formiga, J. K. "Formas normais no estudo da estabilidade para L4 no problema fotogravitacional". ". Phd Thesis, National Institute for Space Research, São José dos Campos, Brazil, 2009.

[6] Zanardi,M. C. "Study of the terms of coupling between rotational and translational motion".Celes. Mech. 39(2), 147 – 164, 1986.

[7] Silva, W. R., Zanardi, M. C., Cabette, R. E. S. and Formiga, J. K. S. Study of stability of rotational motion of spacecraft with canonical variables. Mathematical Problems in Engineering, v. 2012, 12 p., 2012.

[8] Arnold, V. I. "Geometrical methods in the theory of ordinary differential equations". Berlin: Springer- Verlag, p. 334, 1983.

[9] Silva, W. R., "Estudo da estabilidade do movimento rotacional de satélites artificiais com variáveis canônicas". Dissertation, UNESP – Universidade Estadual Paulista, Guaratinguetá, Brazil, 2012.

AUTHOR INDEX